Minerals and Metals in Pre-Modern India

Minerals and Metals in Pre-Modern India

Arun Kumar Biswas

D.K. Printworld (P) Ltd.
New Delhi

Cataloging in Publication Data — DK

Biswas, Arun Kumar, 1934 –

Minerals and metals in pre-modern India.

Includes bibliographical references (p.)

Includes Index.

1. Minerals — India — History. 2. Metals — India — History.
3. Metallurgy — India — History. I. Title.

ISBN 81-246-0183-6

First published in India in 2001

Published and printed by:

D.K. Printworld (P) Ltd.

Regd. office : '*Sri Kunj*', F-52, Bali Nagar

New Delhi - 110 015

Phones : (011) 545-3975; 546-6019; *Fax* : (011) 546-5926

E-mail: dkprint@4mis.com

Preface

OUR earlier book entitled MINERALS AND METALS IN ANCIENT INDIA, submitted to the Indian National Science Academy, was subsequently published in two volumes by D.K. Printworld in 1996. We have been fortunate in receiving several technically competent and overwhelmingly sympathetic and appreciative reviews such as the following:

Paul Craddock, *Journal of Historical Metallurgy Society,* 29, No. 2, 1995, pp. 117-20; Editor, *Indian Science Cruiser,* June 1996; Editor, *Minerals & Metals Review,* Annual 1996, pp. 280-7; Tanuja Pant, *Hindustan Times*, 4 August 1996; C.V. Sundaram, *Current Science*, Vol. 71, No. 7, 10 October 1996, pp. 587-8; Ranabir Chakrabarti, *DESH,* 13 December 1997, pp. 125-9, also 18 April 1998, pp. 13-14; Ajay Mitra Shastri, *Indian Journal of History of Science*, 32 (3), 1997, pp. 297-300, etc.

Craddock acknowledged that the work is 'ambitious, being a survey of the whole range of inorganic materials by a noted materials scientist', but failed to pin-point our original contributions on the deep antiquity of brass and zinc in Lothal and Atrañjikhera, phase studies of ancient zinc slag, etymology of beryl and emerald, etc. The reviewer in the *Minerals and Metals Review Annual 1996* was more specific:

> Professor Arun Kumar Biswas captures a panoramic view of the hoary, richly variegated cultures which, in their final analysis, lead him not only to question the diffusionist theory concerning the 'Āryan intrusion', but also to highlight, among a range of his first-time-arrived conclusions, the primacy of India in the areas of non-ferrous ore mining, production of carburised iron, laminated iron sheets, wootz, steel, forge-welding of wrought iron, distilled zinc and high-zinc brass, etymology of *vaidūrya* and *marakata*, etc. . . .

> In two volumes, the book tells the fascinating, coherently woven story of the minerals and metals — from across the entire sub-continental sprawl of the old world India.

The present volume of MINERALS AND METALS IN PRE-MODERN INDIA may be considered to be the third volume in the series or as a sequel to the earlier two volumes. We are grateful to the reviewers for the several suggestions made for changes in the earlier two volumes which would be incorporated in the next edition. Ranabir Chakrabarti expected discussion on salt in ancient India, and this he would find in the present volume. Whether in the sector of salt or gems, minerals or metals, the Indian sub-continent has witnessed remarkable continuity from the pre-Harappan to the pre-modern eras. This we have meticulously documented in the area of zinc and brass, earning from the reviewers some words of deep appreciation.

The conclusions that we have reached in the present volume have been partly communicated elsewhere. Several papers have been published in *Indian Journal of History of Science* in 1993, 28 (4), pp. 309-30 (primacy of ancient India in zinc and brass), and in 1994 : 29 (2), pp. 139-54 (on etymology of beryl and emerald); 29 (3), pp. 389-420 (gem minerals); pp. 421-63 (non-gem minerals); 29 (4), pp. 579-610 (iron and steel in pre-modern India). A very special article "Minerals and Metals in Medieval India" has been published as chapter 11, pp. 275-313 in *History of Indian Science, Technology and Culture*, Vol. III, Part I edited by Professor A. Rahman, Oxford University Press, 2000.

We have particularly emphasised in the cited articles as well as in this volume, the reasons for medieval decadence of S&T in the sub-continent. Writing on the indigenous industries and traditions in pre-modern India, we have paid our tribute to the indomitable artisans of India representing so many castes, religions, languages and sub-cultures. Their vitality was sustained through centuries, and could not be extinguished either by the Hindu obscurantism, Muslim fundamentalism or British colonialism. Our entire work on *Minerals and Metals in Ancient* and *Pre-Modern India* is indeed dedicated as a tribute to those indomitable artisans who have ceaselessly toiled through the centuries. Their contributions to our civilisation have been as abiding and monumental as those of seers and philosophers.

Abstracts of the Reviews on

"Minerals and Metals in Ancient India"

Vols. I & II

by

Prof. Arun Kumar Biswas

and

Dr. (Mrs.) Sulekha Biswas

"THIS over 800 pages beautifully produced set of two volumes is the first authentic comprehensive work on the theme of such a vital importance. . . . (In volume I) there are a large number of new and somewhat startling-looking conclusions arrived at by Prof. Biswas, which are well-supported by the evidences cited by the author. The second companion volume on the literary evidences is authored jointly by the Biswas couple. We congratulate the authors for accomplishing the gigantic task so very creditably, and the publishers, D.K. Printworld, for bringing it out so very beautifully. Every library specialising in historical, mineralogical and metallurgical literature would do well to have it."

— Prof. Ajoy Mitra Shastri

in *Indian Journal of History of Science*,

32 (3), 1997, pp. 297-300

"The Biswas couple have brought their wide-ranging academic experience and scholarship to the formidable task they had set for themselves of collection, collation, evaluation and construction of a vast body of historical material, to present substantial and valuable conclusions. . . . Notwithstanding the overpowering volume and detail of material presented in the two volumes, what facilitates the reader and sustains the interest is the structured presentation. . . . Maps, chronological presentations, large number of tables, the beautiful colour plates, all add great value to the presentation.

The Biswas couple have produced a very impressive and valuable compendium of information, that should serve as a consolidated sourcebook for future historians and research workers.

— Prof. C.V. Sundaram

in *Current Science*, Vol. 71, No. 7,

10 October 1996, pp. 587-8

"The book, in two volumes, tells the fascinating, coherently woven story of minerals and metals from across the entire sub-continental sprawl of the old world India. . . .

Professor Arun Biswas captures a panoramic view of the hoary, richly variegated cultures — which in their final analysis, lead him not only to question the diffusionist theory concerning the 'Aryan intrusion' but also to highlight, among *a range of his first-time-arrived conclusions*, the primacy of India in the areas of non-ferrous ore mining, production of carburised iron, laminated iron sheets, wootz, steel, forge-welding of wrought iron (as in New Delhi's Iron Pillar), distilled zinc, high zinc brass; besides diamonds and gems of various descriptions, etymology of *vaidūrya* and *marakata*, etc."

— Prof. Murali M. Biswas
in *Indian Science Cruiser*
and *Minerals & Metals Review, Annual* 1996

"The readers would learn from these two giant volumes a lot about mines, minerals and metals in ancient India, a topic on which detailed deliberation has so far been rare and scanty. The Biswas couple have integrated their expertises in applied chemisty, mineral engineering, metallurgy and Sanskrit in an admirable fashion. In the opinion of the reviewer, the most thrilling and significant portion of the book is the treatise on ancient zinc and brass, the mining and metallurgy, the deep mine with staircase, implements and ante-chambers, utensils and lamps and so on. . . ." (translated from Bengali)

—Prof. Ranabir Chakraborti
in the Bengali Periodical DESH,
13 December 1997, pp. 125-9

"One of the most important impressions which persist after perusal of these volumes is the significance of modern scientific data in unambiguous archaeological interpretations. The authors need to be congratulated for this. . . . This is a very lucid, authoritative and widely useful work which will fill a much felt void."

— Tanuja Pant
in *Hindustan Times*, 4 August 1996

Contents

Acknowledgements

We are deeply indebted to the Indian National Commission for History of Science for sponsoring our work on *Minerals and Metals in Pre-Modern India*, extending financial assistance for the period 1991-4, and allowing us to publish the report.

We heartily thank the Commission and the staff of the History of Science Division of the Indian National Science Academy, New Delhi for their splendid and manifold assistance. Dr. Mrs. Sulekha Biswas, Dr. A.K. Bag and Prof. Irfan Habib have provided us assistance and constant encouragement. We express our gratitude to them.

Jan 14, 2001 **Arun Kumar Biswas**

1

Introduction

THE Indian National Science Academy (INSA), New Delhi, had sponsored our research work (1987-91) on MINERALS AND METALS IN ANCIENT INDIA. The final report was submitted in two volumes. The first volume devoted to the travellers' accounts and archaeological evidences contained 21 chapters, 59 tables and 24 figures. The second volume summarised the literary evidences (particularly from the Sanskrit literature) in 10 chapters, 3 tables and 5 useful appendices. D.K. Printworld Pvt. Ltd., New Delhi, has published the monograph in two volumes.

The coverage period of the above-mentioned work spanned from the Pre-Harappan millennia to about AD 1200. We planned to cover the subsequent medieval period in India, and preferred to replace the word 'medieval' by 'pre-modern'.

As in the earlier ancient period, the foreign travellers' accounts were found to be more weighty than the description penned by indigenous writers. Our present work entitled MINERALS AND METALS IN PRE-MODERN INDIA, again sponsored by INSA, New Delhi (1991-4) has been considerably enriched by writings of eminent foreigners/travellers such as Al-Birunī, Amir Khusrau, Jean Baptiste Tavernier, Henry Howard, D. Havart, Alexander Hamilton, George Pearson, Benjamin Heyne, F. Buchanan-Hamilton, H.H. Voysey, James Franklin, H. Yule, J.M. Heath, S.G.T. Heatly, James Prinsep, etc. We are happy that we have been able to fully exploit the classic writings of Valentine Ball whose 1881 treatise on ECONOMIC GEOLOGY (OF INDIA) is a veritable compendium of evidences of *pre-modern* mining, mineral processing and extractive metallurgy in India.

The present monograph has seven chapters. Chapter 2 on gem minerals deals with travellers' accounts, Moghul Treasury, itemised discussions on pearl, coral, chalcedony, beryl, etc., a special section on diamond mining and finally concluding remarks. Appended at the end of the chapter are some Persian/Arabic/Hindustani terms for gem minerals used in medieval India.

Non-gem minerals in pre-modern India would be discussed in chapter 3. Principal items therein are salt, alkaline salt or *reh*, saltpetre, metallic ores — of copper, lead and zinc, special minerals like *sehta* or ferro-cobaltite, sulphates of copper, aluminium and iron, gold-bearing river sand, building stones and iron ores of different varieties in different geographical locations of India.

Chapter 4 on iron and steel in pre-modern India emphasises diversities of ores processed, smelting techniques and products in Southern, Central, Eastern and Western sectors of the country. Iron-carbon alloys, mentioned in *Rasa-Ratna-Samuccaya*, and wootz (crucible steel)-making in south India have been specially emphasised. The chapter includes detailed discussions on three different types of iron and steel enterprises in pre-modern India, charcoal crisis and other reasons for stagnation and eventual extinction of indigenous iron industries.

Chapter 5 summarises the primacy of India in brass and zinc metallurgy ever since the Lothal and Atrañjikhera eras, and the singular achievements of medieval Zāwar in the art of zinc distillation. It is established that there was a technology-transfer of the zinc downward distillation technology from Zāwar, Rajasthan in India to Bristol, England sometime during the early eighteenth century.

Various traditions of the metallic art works in pre-modern India would be summarised in chapter 6. One outstanding example is the Bidriware of Bidār based on high zinc low copper content alloy provided with a black patina and encrusted with gold or silver wire. This belonged to a broader family of damascene art products probably introduced by the immigrants from Syria or Iraq. Other traditions which survived the ravages of time are related to brassware, gold and silverware, tin-coating of metallic utensils, *mīnākārī* or the art of enamelling, jewel-setting, etc. These traditions developed 'without the use of steam or machinery' have survived on account of the artistic sense and perceptive originality of artisans. On the other hand, the traditions of indigenous mining and smelting dwindled to extinction on account of lack of intellectual support from the educated community and the ruling class, numerous technical problems, lack of innovation and the ruthless emerging domination of the colonial power.

Chapter 7, the last one in this monograph, contains a brief summary and concluding remarks. The positive achievements in the pre-modern Indian industries are highlighted. During this period India was receiving impetus in the area of material technologies from the emerging colonial powers, often imperceptibly; this aspect has not been fully researched so far. At the same time, there were technology transfers from India to the West, specially in the metallurgical sectors of wootz and distilled zinc. The chapter ends up with the articulation of causes behind decline of the indigenous industries, the superb vitality of the artisans' traditions in India and lastly the utility of the historical studies such as the one described in this monograph.

2

Gem Minerals in Pre-Modern India

A COMPREHENSIVE report on minerals and metals — including gem minerals in ancient India was prepared and submitted to the Indian National Science Academy (INSA).[1] A summary on gems and minerals[2] included the nature of evidences about the antiquity of their uses, narrative from Mehargarh to Taxila, specific discussions on some selected gem-minerals (and metalliferrous ores) and lastly the need for further studies and research on the subject.

It was emphasised that the subject is a continuing one in the annals of history, and the demarcation line between the 'ancient' and 'medieval' period in India (AD 1200 is the conventional date) is bound to be arbitrary. Instead of 'medieval' we prefer the term 'pre-modern' to indicate the nineteenth-century vantage point looking backwards to find out what India had achieved in the field of gem minerals during the period sixth to eighteenth century.

This chapter is divided into the following sections (a) travellers' accounts of Indian gems; (b) the gem treasury of the Moghuls; (c) itemised discussions on specific gem minerals such as pearl, coral, crystalline and chalcedonic quartz, agate and carnelian, corundum, ruby and sapphire, zircon, garnet and beryl, etc.; (d) a special treatise on Indian diamond mines in the pre-modern era; and lastly (e) concluding remarks.

Travellers' accounts of Indian gems

India's traditions in ancient and medieval gems are authenticated not so much by archaeological evidences as by the travellers' accounts. Valentine Ball has collated some of these accounts.[3-5]

Pliny in his *Historis Naturalis* (AD 77) had given an extraordinary amount of information regarding precious stones and metals, a large proportion of them being of Indian origin. He referred to Indian *adamas* (diamond), *smaragdus* (emerald), beryl, opal, etc. *The Periplus of the Erythrean Sea* (AD 80-89?) mentioned the export of Murrha, the cups and vases of carnelian and agate. The sardonyx mines of Ptolemy (AD 140-60) are probably identical with the famous carnelian and agate mines of Rājpīpla or Ratanpur. Ptolemy referred to diamond mining on the Adamas river, the location of which has been widely debated. Colonel Yule identified this with Brāhmaṇi in Orissa, but it could be Mahānadī.

During Emperor Justinian's time (sixth century AD), the Persians became rivals of the Romans at the Indian ports, and this has been described by Cosmos, an Egyptian merchant. During the advent of Islam, the Arabs wrested the Oriental trade of gems from the Persians. One of the contemporary Arabic works bears the title: *Meadows of Gold and Mines of Jewels*.

The outstanding thirteenth-century Persian work on precious stones by Mahomed Ben Mansur was translated into French by Joseph Von Hammer (*Mines de l' orient*, Vol. VI), an English version of which was published (*Asiatic Journal*, Vol. IX, 1820, p. 345). Although Ben Mansur mentioned seven varieties of diamond, his knowledge regarding the geographical location of the Indian mines was very vague : "it is found probably in the *yāqūt* mines of the eastern India". He described six varieties of *yāqūt* gems and many sub-divisions including sapphire, ruby, emerald, topaz, etc., emphasising on their hardness. Ben Mansur described how an earthquake, during the time of the Caliphate of the Abbasides, led to the discovery of balas ruby mine in Badakshan embedded in white limestone matrix. He also referred to the popular usages of turquoise, rock crystal ('which could be melted like glass and coloured so as to imitate *yākūt*, *lāl* or emerald'), amethyst, emerald, jade, jasper, etc. Ben Mansur described Abu Rihan Al-biruni's water-displacement experiments to determine specific gravities of precious gems.[6]

Al-beruni had written about Indian gems, and by the eleventh century when the treasures of north India were either plundered by the Muslims or taken to the south for safety, the empires and temples of south India became store-houses for precious metals and gems.

The Muslim loot was extended to south India. Hazrat Amir Khusrau accompanied Malik Nāib, Alāuddīn Khaljī's army general in AD1310-12 military expedition to south India and described the colossal treasure of gold, emerald and other gems which had been collected during the earlier periods of ancient India.[7] When the fort of Wārangal was attacked, Rāi Laḍḍar Deo (Pratāpa Rudra II) had to give up his treasures and jewels. Khusrau described (Ref. No. 7, pp. 76-7):

> The boxes carried by the elephants were full of valuables and gems, the excellence of which drove the onlookers mad. Every emerald (*zabarzad*) sparkled in the light of the sun. . . . The corundum/sapphire (*yāqūt*) dazzled the eye in the sun. The cat's eye (*ainul hirrāt*) and the cock's eye (*ainul dīk*) were so brilliant. The lustre of the rubies (*lāl*) illuminated the darkness of the night. The emeralds had a fineness of water that could eclipse the lawn of paradise. The diamonds (*ilmā*s) would have penetrated into an iron heart like an arrow of steel. The other stones were such that the sun blushed to look at them. As for the pearls, you would not find the like of them, even if you kept diving into the sea through all eternity. The gold was like the full moon of the twelfth night; it seemed that in order to ripen it, the *alchemist* sun, had lighted its fire, and the morning had blown its breath, for years. . . . The Arīz-Mumālik (gemmologist) divided the jewels into 'genus' and 'species', 'class' after 'class', and had everything written down. . . . Among them was a jewel (Koh-ī-Nūr ?), unparalleled in the whole world.

The sack of the golden temple of Barmatpur was similarly described (Ref. no.7, pp. 102-7):

> Its roofs and walls were inlaid with sparkling rubies and emeralds, and after gazing at them, red and yellow spots came before the spectator's eye. . . . The heads of the idol-worshippers came dancing from their necks. The golden bricks rolled down and brought with them the plaster of sandal wood; the yellow gold became red with blood, and the white sandal turned scarlet. The foundations of the temple, which were mines of gold, were dug up, and its jewelled walls, which were mines of precious stones, pulled down. . . . There were five hundred *man*s of precious stones.

Malik Nāib reached Delhi in AD 1312 with 612 elephants, 20,000 horses, 96,000 *man*s of gold (the figure seems absurd), many chests of jewels and pearls. The old men of Delhi declared:

> "No one remembers such treasures and spoils brought ever to Delhi."
> — Ref. No.7, p. 117

Alāuddīn Khaljī's loot has been authenticated by the Jaina scholar Ṭhakkura Pherū born around AD 1270, who was probably an assay-master in Alāuddīn's court.[8] Pherū 'saw with his own eyes the vast collection of gems — like the ocean — in the treasury of Alāuddīn — the emperor of the *kali* age' and also witnessed the examination of the gems by the experts.[8] (*Rayana-parikkha*, 1.4-5)

In his book, Pherū followed the framework of Indian *ratnaśāstra*s or gemmological literature. However, he described new varieties and sources of ruby, sapphire, emerald, etc. The contemporary gem trade and tariff prices were described. In a separate section (7.103-105), Badakshan spinel or balas-ruby, Yemen carnelian (*akīk*) of deep red hue and the beautiful blue turquoise of Niśāpur and Al-Moussul were mentioned as gems imported from Persia.

Ibn al-Afkānī, an Egyptian (who died on AD1348) and a contemporary of Pherū wrote in his Arabic work on gemmology *Nachb al.* Gawahir, about the Indians' preference for white and yellow diamonds, the beautiful diamond specimens in the court of Qutbuddīn Mubārak (AD 1316-20) and the seven varieties of *yāqūt* or ruby — the best *rummānī* variety having the colour of the fresh seed of pomegranate. Al Ta' ālibī enumerated the best contemporary sources of different gems: turquoise of Nīśāpur, ruby and sapphire of Sri Lanka, carnelian of Yemen, garnet of Balkh and the spinel of Badakshan.[9]

Marco Polo, the famous Venetian traveller of the thirteenth century, reported diamond trade going on through the ports (such as Motupalle) of the Gunṭūr district, the big stones going to the Indian kings and the great Khan, and 'the refuse of the finer stones to Europe'. After the rains the beds of torrents were searched for diamond, wrote Marco Polo.

He referred to other precious stones, the trader of Balas rubies and lapis lazuli of Badakshan. Enormous amounts of gold were being stored by the south Indian princes. Copper, gold and silver were being imported into Malabar and Cambay from the eastern countries of Myanmar, Malaysia, Indonesia, etc. These treasures in their turn were plundered by the Muslim invaders from the northern India.

Muhammad Bin Tughlak loaded hundreds of elephants with the precious spoils of Hindu temples.

Ibn Battūtā wrote in his *Travels in Asia and Africa 1325-1354*, about pearl fishing of incredible magnitude going on at Rāmnāḍ (Adam's bridge) in Tamil Nadu and also sapphire mining in Sri Lanka. Ferishta described in 1425 the long-deserted diamond mines in Madhya Pradesh probably those of Wairagarh — and the great wealth possessed by the kings of south India in the form of precious stones and bullion. For two centuries, there was a resurgence of Hindu imperial culture at Vijayanagara (AD 1350-1564), the economic history of which has been narrated by Burton Stein (Ray Chaudhuri and Habib[10a]). The gold and diamond treasures of the Vijayanagara empire were described by the Venetian Nicolo Conti (AD 1420) mentioning washing procedures for diamonds executed in the bed of rivers.

Odarodo Barbosa reported in 1519 the gem mart at Vijayanagara transacting jewels — from Myanmar, Sri Lanka, lower Kr̥ṣṇā valley — such as rubies, spinel, diamond, sapphire, topaz, turquoise, hyacinth, emerald, etc. Barbosa reported the sources and full account of the values of these gems. He also described how the foreign Muslims (Moors) were conducting pearl-fishery and trade in Calicut and other parts of Malabar 'without interference from the king'. Gradually they faced competition from the Portuguese and the Dutch.

In the year 1563, Garcia da Orta, in his famous work *Simples and Drugs of India* written in Portuguese mentioned four large Indian diamonds, one of which had the 'size of a small hen's egg' and was kept at Vijayanagara. The others weighed 150, 175, and 312 carats. Garcia da Orta mentioned that at Vijayanagara the big gems weighing more than 30 mangelis (750 carats?) were handed over to the king. Even the tailings were rewashed to recover additional quantities of diamond. The first discovery of diamond in Brazil was reported at this time although commercial processing in that country started around AD 1728 ending the Indian monopoly in this gem. Hyacinth and garnet were found in Calicut and Cannanore; garnet being distributed through the whole of Cambay and Bālāghāṭ, the latter also produced imitation emerald made of glass.

In Golkoṇḍā, Qutbśāhī kings used to collect excellent workmen, and a number of workers on precious stones and kept them busy at the *kārkhānā*s to do any work for anybody else.[10b]

Ralph Fitch and his companion Newberry left an account in AD 1583 about the gem mart at Belgaum dealing with diamond, ruby, sapphire and many other precious stones. Their companion, one William Leades, a jeweller, stayed behind at Cambay to serve the local king. In 1590, Abul Fazl, the author of *Āin-ī-Akbarī*, wrote on the diamond mine of Wairagarh and gold-washing in the rivers in Kashmir.

Benedict Goez reported in 1602 the production of jade in Kashgar, 'a certain shining marble, which for want of a fitter name, Europeans call jasper'. Vessels, ornaments made of jade were imported by Moghuls in India. Nephrite, jadeite and chloromelanite — all the three varieties of jade available in the regions of Kashgar, Yarkand and Khoṭan, used to be processed by early Buddhists of Central Asia. Later, this trade flourished under Ulugh Beg in central Asia. Jehāngīr became a great patron of Timurid workmanship in jade during the early part of the seventeenth century.[11]

In 1602-3, the Portuguese traders took away from India not only indigo, pepper and other spices but also some quantities of carnelian (*laqueca*) and precious stones like diamonds and pearls. During 1618-19 the English trade from Surat showed purchases of precious stones from India (bulk of it was of course indigo and calico). The Dutch traders also purchased some Indian (Golkoṇḍā) diamonds in the 1620s. (These data have been provided by W.H. Moreland in his treatise of economic history 'From Akbar to Aurangzeb'.)

De Boot's work in 1609 and John Ogilby's writing in 1673 recorded some valuable information about Indian gems and mineral resources, but their data were tentative and not always reliable. Tavernier (1665-69) mentioned 'excellent gold' from rivers in Kashmir and 'not so good' gold and also silver coming from Assam. His notes on tin ores from Malaysia, ruby from Myanmar and sapphire from Sri Lanka did not provide accurate details about their sources.

Tavernier described Goa as the place where *formerly* (before his time), 'there was the largest trade in all Asia in diamonds, rubies, sapphires, topazes and other stones — there was also at Goa a large trade in pearls'. The miners and merchants used to go there, eluding their kings, to fetch the best prices for their best gems (Ref. no. 5, p. 95). The Portuguese monopoly of the international trade in Goa declined during the last decade of the sixteenth century with the ascendancy of the marine power of the Dutch.

The gem treasury of the Moghuls

The Moghul dynasty consolidated the gem treasures of the Paṭhān rulers. It is said that 'Bābar's diamond' was received (AD 1526) by his son Humāyūn from the family of Rājā Bikramjīt, when he took possession of Agra. It had been earlier acquired from the Rājā of Mālwā by Alāuddīn in the year AD1304. Erskine and King have identified Bābar's diamond, weighing according to Ferishta 8 *miṣkāl* or 186 English carats, to be identical with the famous *Koh-ī-Nūr* now in the British throne. Valentine Ball however refuted this identification,[5] and suggested that Bābar's diamond is 186 carat, flat *Daryā-ī-Nūr* now in the Shāh's treasury in Teherān.

Akbar was the first Moghul who organised a 'treasury for precious stones' as described by Abul Fazl (*Ain* 3). The foundation of the treasury rested on 'the four pillars' of an intelligent, trustworthy treasurer, a zealous *dārōghā*, a band of skilful jewellers and a retinue of assistants. Rubies, diamonds, emeralds, red and blue *yāqūt*s were categorised under 12 classes and pearls into 16 classes. Values of jewels and charges for boring pearls were recorded.[12]

Jehāngīr was a great lover of gems, particularly diamonds and jades. By the time of Shāh Jahān, the treasury had a huge stock of diamond, emerald, lapis lazuli, ruby (some inscribed), sapphire and also rosary, necklace and ornament studded with them.[13] After imprisoning his father, Auraṅgzeb wanted to possess the entire collection, when Shāh Jahān sent back the word that he would pound all the gems in a pestle and mortar rather than part with them. Auraṅgzeb ascended the throne with only one oriental topaz on his cap, and had to wait for his father's death before gaining the entire treasure of jewels.

Tavernier was allowed to examine Auraṅgzeb's jewels on the 2nd of November 1665. Being a jeweller and gemmologist himself, his account is of great importance.

The first piece that he was allowed to examine in his hands was the 'Great Moghul' diamond, a round rose, very high at one side. When Mir Jumla, who betrayed the King of Golkoṇḍā, his master, presented this stone to Shāh Jahān, the stone weighed 787 carats. It had been discovered in the Kollūr mine in 1656. The wasteful grinding treatment to it subjected by a Venetian named Hortensio

Borgio, reduced it to 268 English carats and this is how Tarvernier found the piece in Auraṅgzeb's collection.

Tavernier described and drew the shapes of this 'Great Moghul' and seven other pieces which he saw in Auraṅgzeb's treasury (Ref. No. 5, pp. 96-9). Their weights were recorded by him. One of them weighed (137 carats) and looked like the Austrian Grand Duke of Tuscany's diamond. Another one, a 230 carat table diamond, had been seen by Tavernier once before at Golkoṇḍā (AD 1642) 'in the possession of merchants', and valued by him at 5 lakh of rupees. (Vide Plate 1, p. 197)

Tavernier also saw a jewel set with 12 diamonds, the central one being 'a heart-shaped rose of good water'. There was another jewel set with 17 diamonds.

Auraṅgzeb had a large collection of pearls, the largest being a pear-shaped one, a little flattened on both sides, weighing 70 *rattī*s or approximately 60 carats.

There were two chains, one of pearls and rubies of different shapes pierced like the pearls; the other of pearls and emeralds, round and bored. All the pearls were round and from 10 to 12 *rattī*s each in weight. Tavernier described:

> In the middle of the chain of rubies, there is a large emerald of the old rock (perfectly crystalline) cut into a rectangle and of high colour, but with many flaws. It weighs about 30 *rattī*s. In the middle of the chains of emeralds there is an oriental amethyst (actually purple sapphire), a long table, weighing about 40 *rattī*s and the perfection of beauty.
>
> — Ref. No.5, p. 318

Tavernier also described the 158 carat oriental topaz (actually yellow sapphire) of octagonal shape which Auraṅgzeb wore on his cap during coronation, several cabuchons (polished but not cleaved) of ruby and balas ruby, etc. There was a ruby square shaped of two inches dimension, bearing the name of Jehāngīr which was taken to Persia and later ended up in Ranjīt Singh's collection.

Auraṅgzeb was the proud possessor of all the gold and jewels worth 4 million pounds owned by Dārā Shikoh and a larger amount of treasure possessed by his father. In addition, he had seven magnificent thrones, one wholly covered with diamonds, the others with rubies, emeralds or pearls.

THE PEACOCK THRONE

The principal throne was 6 feet long and 4 feet wide. The four bars which supported the base of the throne were inlaid with gold and enriched with numerous diamonds, rubies and emeralds. The middle of each bar was decorated with ornamental square cross constituting of either one central ruby with four emeralds round it, or one central emerald with four rubies on four sides. The intervals between rubies and emeralds were covered with diamonds. There were similar decorations all around the throne. Tavernier counted 108 rubies, all cabuchons, weighing 100 to 200 carats each and 110 emeralds weighing 30-60 carats each on the great throne. Tavernier wrote:

> The underside of the canopy is covered with diamonds and pearls, with a fringe of pearl all round, and above the canopy, which is a quadrangular-shaped dome, there is a peacock with elevated tail made of blue sapphires and other coloured stones. The body of the peacock is made of gold inlaid with precious stones, having a large ruby in front of the breast, whence hangs a pear-shaped pearl of 50 carats or thereabouts, and of a somewhat yellow water. On both sides of the peacock there is a large bouquet of the same height as the bird, consisting of many kinds of flowers, made of gold inlaid with precious stones. . . .
>
> In my opinion the most costly point about this magnificent throne is that the twelve columns supporting the canopy are surrounded with beautiful rows of pearls, which are round and of fine water, and weigh from 6 to 10 carats each. At a distance from the throne, the two umbrellas fixed on either side are covered with diamonds, rubies and pearls.

The cost of the famous throne has been variously estimated as Rs. 4 crore (Bernier) to Rs. 10 crore (Tavernier). The throne was taken to Persia by Nādir Shāh. Lord Curzon believed that the Peacock Throne, now in Teherān, priced at 2.6 million pounds is a re-construct on utilising the original materials.

Nādir Shāh also took away in 1739, the entire (70-80 million pounds worth) gem treasure at Delhi, including the 268 English carat 'Great Moghul' diamond. He named the celebrated piece as *Koh-ī-Nūr* or the Mountain of Light. When this gem reached Ranjīt Singh of Punjab in 1813, through a long chain of violent events, it weighed 185 carats. Valentine Ball has explained this discrepancy of 83 carats in terms of mutilations sometime during the period 1739-1813. When the gem was taken from Lahore to England in 1849, Mr. Tennant, a gemmologist observed in the piece

> two large cleavage planes, one of which had not even been polished, and had been distinctly produced by fracture.
>
> — Ref. No. 5, pp. 342-3

Exactly a century after Nādir Shāh's plunder of Delhi, occurred the demise of Ranjīt Singh. On his deathbed, he expressed a wish that the diamond, then valued at one million sterling, should be sent to the Jagannāth (temple) at Purī, but this intention was not carried out. Through Delhi, Teherān and London it has remained as the property of the usurpers.

The huge collections of gems and jewelleries, looted from India, adorn the museums of London and Teherān. The chests 'filled with gold, silver, diamonds, pearls and emeralds from the Moghul's treasury' are now in the Teheran Museum. This has been characterised by a Curator of the Smithsonian Institution as 'perhaps the greatest jewel treasury of all times'.

When Timūr plundered Delhi in 1398-9, he took away many gems including a ruby which was later celebrated as Timūr's ruby. It came back to Jehāngīr, a descendant of Timūr, was again taken away by Nādir Shāh, back to Ranjīt Singh and finally to the Crown in England. Along with *Koh-ī-Nūr* and *Daryā-ī-Nūr* diamonds, Timūr's ruby are now the possessions of the latest plunderer of the Indian jewels, the British royalty.

Discussions on specific gem minerals

Let us now record some of the pre-modern or medieval observations on specific gems in India, avoiding both the treatise on the ancient period which we have done before extensively,[1] as well as the modern or the twentieth century scenario. First, we would take up two soft gem materials which are of organic nature and hence non-mineral origin: pearl and coral.

The pearl fisheries in India flourished early in the ancient period. In seventeenth century, Tavernier reported about the trade going on in the Gulf of

Persia as well as in the gulf separating Sri Lanka and south India. He observed that the oyesters of the Manār Strait open up themselves, spontaneously, 'five or six days sooner than those of the Gulf of Persia, because the heat is much greater at Manār'. The latitudes in the gulf between Sri Lanka and India are between 8-9° whereas that in the island of Bahrein is 27°. Tavernier also corrected the earlier held, erroneous impressions 'that the pearl originates from the dew of heaven, and that but one is found in each oyster'.

The pearl-fishing in the Manār used to take place twice a year during March-April and then August-September ('the heavier the rainfall, the better the pearl-fishery'), and the sale lasted from June to November. Tavernier wrote:

> It should not be supposed that a great profit is earned by those who fish for pearls; for if the poor people who engage in it had anything else to do, they would leave the fishing, which merely saves them from dying of hunger. . . . Those who fish like to know beforehand whether it will pay. . . The people of Manār are better fishers, and remain for a longer time under water than those of Bahrein; they do not place any clips on their noses nor cotton in their ears to keep the water from entering, as is done in the Persian Gulf. — Ref. no. 5, pp. 91-4

The pearls at Manār usually 2-4 carats in weight, were 'the most beautiful both as regards water and roundness, of all the fisheries'.

The Portuguese 'used to protect' the poor fishermen of Manār from the Malabaris, but that was before Tavernier's time, when the Dutch displaced even the Portuguese and extracted not only the 'protection levy' but also one day's catch (in a week) from the fishermen. The large trade of pearl and other precious gems at the Portuguese Goa had become, during Tavarnier's time a thing of the past.

The pearl trade prospered under the Dutch. The Indians liked the round as well as the 'baroque' (irregularly shaped) pearls. The sketch of the largest pearl which Tavernier saw and drew (the 50-carat, yellowish, pear-shaped one, hanging from the neck of the peacock in the celebrated throne) looks like an oblong hanging water droplet. Finally, Tavernier made the following interesting comment:

> All the Orientals are very much of our taste in matters of whiteness, and I have always remarked that they prefer the whitest pearls, the whitest diamonds, the whitest bread, and the whitest women!
>
> — Ref. no. 5, p. 91

Tavernier gave some details about coral fishing and trade during his time. He mentioned seven areas in the Mediterranean Sea region where coral fishing was being done: three areas around the coast of Sardinia-Alghero, Boza, St. Pierre; fourth, the island of Corsica, coast of Sicily near Trepani, the coast of Algiers and the seventh, the north coast of Tunis. Tavernier noted that whereas coral did not rank among precious stones in Europe, the people in the East preferred this, 'one of the most beautiful of Nature's productions', to precious stones. Tavernier observed :

> It is much to be wondered that the Japanese give so much money for a fine piece of coral. . . . The common people wear it and use it as an ornament for the neck and arm throughout Asia, but principally towards the north in the territories of the Great Moghul (Auraṅgzeb), and beyond them, in the mountains, of the Kingdoms of Assam and Bhutan".
>
> — Ref. No. 5, pp. 104-7

A small town named Lantegree in Maharashtra was a great centre of coral-polishing in the early seventeenth century. The preference of the Indians and other Asians for coral was manifest even during the ancient period, and the *reason* for this preference has been subject of many dissertations. Many travellers like Pliny and Marco Polo attributed the preference for coral 'to the way its tints adapt themselves to set off a dark skin (of the Asians)'. The real reason could be a religious antecedant. The reddish yellow coral is known as *rudrākṣa*, the eye of the Śiva and a symbol of renunciation and spirituality. It must have gained popularity during the Tāntric period of Hinduism and (Mahāyāna) Buddhism. The 'ornament for the neck' used by the common people was a rosary of coral beads, which were counted during prayers.

The Indian tradition of jewelleries made of rock crystal is very ancient but not well-documented. The present author has suggested[14] that the south Indian generic term *veḷ-ur* stood for white crystal town and gave rise to the more popular terms of *veluriya* or beryl and *veloari* or rock crystal.

There was widespread use of chalcedonic and crystalline quartz (in ancient India), which were available in the rocks near Tanjore and the estuaries of Kṛṣṇā and Godāvarī. Ball reported (Ref. no. 5, p. 502) that the lapidaries at Vellum (latitude 10° 43' and longitude 79° 12') had skilled workers in different varieties of rock crystal, such as the ordinary pellucid quartz, smoky quartz, cairngorum and amethyst. The last was brought from Kangiam in Coimbatore, and the rest derived from Cuḍḍalur (tertiary) conglomerates. The ornaments made were chiefly of broach stones cut in the brilliant, rose and other patterns. Godāvarī district, Hyderābād state and Sambalpur district of the Central Provinces also provided brilliant rock crystals.

Auraṅgpur of the Gurgāon district, 15 miles south of Delhi, had Arāvalī quartzites from which quartz crystals were extracted, and these might have been used for making vases and ornaments. Tavernier saw Auraṅgzeb drink from a large cup of rock crystal placed on a golden saucer, enriched with diamond, rubies and emeralds. What a sight ! After the 1857 War of Independence, the Delhi Palace was looted and found to contain many drinking vessels, vases and pitchers made of rock crystal, which were later described by Valentine Ball.

Chalcedony is known as a compact of silica and hydrated silica, slightly lighter and softer than quartz. Its many varieties are available in India: carnelian (red), sard (brown), plasma (dark green), heliotrope or blood-stone (green spotted with red), agate and onyx (banded structure), etc. Agate and carnelian were mined, processed and used in India even in the Pre-Harappan times. Yellowish varieties, containing hydrated ferric oxide, used to be dehydrated on heating, and thus reddened to the anhydrous ferric oxide stage and beautiful carnelian in an age-old practice. The best known deposits are found in the Rājpīpla hills at Ratanpur, on the lower Narmadā river. Deposits of carnelian were mined and processed also near the Mahi river, north of Baroda.

The mining and working of the stones at Ratanpur were vividly described in 1878 by J.M. Campbell (*Bombay Gazetteer*, Vol. VI, p. 205) and reproduced by Valentine Ball (Ref. no. 4, pp. 507-13). A suitable extract is given below:

> The mine shafts are about 4 feet in diameter, and on an average about 30 feet deep. At the foot of each shaft, galleries 5 feet high and 4 feet wide branch off on all sides. These passages, seldom more than 100 yards long, in

many cases join the galleries of other mines. At the mine mouth, the stones are chipped, and the likely ones carried to Ratanpur, the village of gems.

The contractor divides the stones into two classes, those which should, and those which should not be baked. Three kinds are *not* baked: onyx called *mora* or *bawa ghori*, the cat's eye called *cheshamdar* or *dola*, and yellow half-clear pebble called *rorī* or *lasania*. The other kinds of pebbles are baked to bring out their colours. During the hot season, the stones are spread in the sun in an open field. During other seasons, the stones are gathered in earthen pots and heated in piles of ignited cowdung cakes. . . . Then the stones are taken to Broach and Cambay.

By exposure to sun and fire, maize yellow gains a rosy tint, orange is intensified into red; the hue of the red carnelian varies from the palest flesh to the deepest blood red. . . . The larger and thicker the stone, the more it is esteemed. Four kinds of agates, the common, the moss, the *kapadvañj*, and the veined, rank next to the Rājpīpla carnelians.

The moss agate, *suabhaji*, comes from Budkotra, 3 miles from Tankara in Morvi. *Kapadvañj* provides another variety of agate. The most valued Cambay agate, the veined agate, *doradar*, comes from Ranpur in Ahmedābād.

Of other stones processed in Cambay, jasper or bloodstone, the chocolate stone, a variegated pebble known as *maimariam* and the rock crystal are found in Gujarat. The lapis lazuli, obsidian and the blue stone *pirojā* (turquoise) are foreign stones brought from Bombay. The rock crystal comes from Tankara in Morvi, the best varieties from Madras, Ceylon and China.

The rough stone generally passes through three processes, sawing, chiselling and polishing (these were described in detail.). . . Cambay agate ornaments belonged to three classes: those suited for the Chinese, Arab, and the European markets.

The trade in Cambay stones at present (1878) supports about 600 families of skilled workmen, and from 500 to 600 unskilled labourers. . .

> . Each process is carried on a distinct workshop under a *kārkhānāwālā* or head of the factory. In each branch of the craft, the heads of factories form a distinct guild or *pañcāyat*. The *akīkiā*, or the agate-dealer must be a man of some capital to buy the incoming rough stones, process them and pass on the finished articles to the agate merchants of Bombay, and through them to Calcutta, China or Jedda.

Ratanpur in the territory of the Rājā of Rājpīpla has been the centre of the 2000 years old (may be older) international trade on articles made of agate and carnelian.

CORUNDUM, RUBY AND SAPPHIRE

Valentine Ball (Ref. no. 4, chapter 9) gave a fairly detailed account of the aluminium-containing gem minerals as found and used in pre-modern India. Corundum, when pure, is sesquioxide of aluminium: Al_2O_3, containing traces of lime, silica and magnesia. It is next to diamond in hardness, and widely known as emery (abrasive) in granular form. Typical, interstitial and lattice-replacing atoms make corundum red, when it is known as 'ruby', and blue, when it is called 'sapphire'.

It is well- accepted that the use of white corundum started in India. Even the name is derived from the Sanskrit word *kuruvinda*. The British travellers of the early nineteenth century reported indigenous mining works on corundum in India. Captain Newbold for example, found in the 1840s widespread corundum mining in the Salem district: at Caranel, Anpore, Mallapollaye and at various localities up the river Kāverī, as far as Coacorambadi, 'where it had been dug up in the fields, and the remains of the excavations were still to be traced'. Newbold also described the mines at Golhushully and Kulkairi in Mysore (*Journal of Royal Asiatic Society*, Vol. VII, pp. 219-24). The corundum occurred in both the crystalline and amorphous conditions in more or less decomposed talcose schists. The material mined from shallow pits was separated into four classes: red, white, scraps of both, and refuse. The first three were conveyed to Mangalore and Tellicherry and were sold to Bombay and Arab merchants for Rs. 12-30 per *kandy* (= 500 lbs?) according to quality. These mines had been opened in 1829.

The stone known as *kurund* used to be mined during the 1850s in the upper parts of the upper Godāvarī district, specially near Bhadrachalam.

The first mention of the occurrence of corundum at Pipra, Singrowli (latitude 23° 57', longitude 82°44') in the erstwhile Rewah State was made by Francis Buchanan Hamilton (*Edin. Phil. Jour.*, Vol. II, p. 305, 1820). He had been prevented from visiting the mine in 1814. Captain W.S. Sherwill's report in 1845 (*Journal Asiatic Society Bengal*, Vol. XIV, Proc., p. 15) indicated that the material at Pipra was of six varieties: *gulābī* rose colour, *massūriā*, like *masūr dāl*, a pea or lentil, *bhaka*, of varied colours, *teliā*, of the colour of the *telī* seed, mica-mixed and fibrolite-mixed. In his papers (*Records Geological Survey of India*, Vol. V, p. 20, 1872 and Vol. VI, p. 43, 1873), F.R. Mallet published accounts of his visit to this mine situated on a hill between Pipra and Kadopani. Several yards (at places 30 yards) thick reddish grey bed of corundum rested between quartz schist and porphyritic gneiss with hornblende rock. Traceable about half a mile, the deposit appeared to Mallet to be 'practically inexhaustible'. Considerable amount of pre-modern mining must have taken place before 1814, and it was still going on in 1871.

Holland described[15] the Indian lapidary (*begri*) using different kinds of discs (*sān*) for cutting precious stones. The corundum-*sān* or lapidaries' wheel as used in the nineteenth-century India, was composed of about two-thirds of finely crushed corundum cemented with one-third of lac resin. Corundum discs of different grades were used for the rough cutting of minerals softer than diamonds, while polishing was done on discs of bell-metal or pewter. Besides the ordinary *begri* or lapidary, there were men whose business was to bore holes through precious stones, and they were known as *bidhiyā*. W. Hoey made a comprehensive report in 1880 on the trade and manufactures of gemstones including rubies and sapphires.

Rubies were reported in the Salem district and the Mahānadī river between Cuttack and Sambalpur, but most materials came from upper Myanmar — places like Kyatpyen, 70 miles north-east of Mandalay. Sri Lanka provided most of the sapphires. Many of the famous rubies known in Europe can be recognised to be of Indian origin on account of the way in which they are pierced through the middle. One such huge specimen acquired in 1867 is now placed in front of the crown of the British queen.[15] Similarly, one Indian sapphire weighing 225 carats was brought to England in 1856.

On 563 carat 'Star Sapphire of India' is displayed in the American Museum of Natural History, New York. The Smithsonian Institution in Washington possesses 'Star of Bombay' corundum, 'Bismarck Sapphire' of Sri Lanka and the Roser Reeves Collection of ruby gems.

Writing sometime in the middle of the seventeenth century, Nicols mentioned about zircon gems being found in Cannanor and Cambay in India.[16] The alluvium at Ellorā contained large obtuse octahedron crystals of zircon along with corundum and these (Sanskrit, *gomed*, red ones known as hyacinth) must have been utilised by jewellers of Cambay.

Garnet jewels were also popular in India. Voysey, Newbold and others reported recovery of garnets from the Mahānadī bed in Orissa, Koṇḍapilli (lat. 16°38', long. 80°36') in the Godāvarī district, Gharibpeṭh, 8 miles south of Paloncha in Hyderābād, etc. There were extensive mines of garnet in the Kiṣengarh State of Rājpūtānā near Sarwar (lat 26°4', long, 75°4'30") from which gems of large size and good colour were obtained, and the Rājā derived a large revenue.

> For a distance of about a mile, the outcrop has been burrowed into. It is believed that the large polished garnets cut *en cabuchon* which are available for sale in principal towns of Punjab, have been obtained in these mines. (Ref. no. 4, p. 523)

The mining and widespread trade of beryl in India, ever since the days of Pāṇini, has been discussed elsewhere.[14] When Britishers came to India, they found that the mines had been nearly exhausted. Clarke reported Mysore beryls (*Madras Journal of Lit. and Sci.*, Vol. IX, p. 121, 1839), and Capt. Newbold wrote on the Coimbatore deposit (*Edin. Phil. Jour.*, 1840). The Paddoor mine in Coimbatore had been worked for two years in 1818 by one Mr. Heath; when Newbold visited, the mining had ceased. The rock there is a vein or dyke of coarse granite consisting of quartz, garnet, mica, etc.; the cavities in cleavlandite felspar contained beryl crystals. Newbold also alluded to a report that beryl used to be mined at Vaniāmbaḍi, at the northen base of the Nīlgiris.

Aquamarine crystals from Kaṅgiam in Coimbatore were exhibited at the Vienna exhibition, during the middle of the nineteenth century. There are current reports during the end of the twentieth century, that many aquamarine crystals may be recovered from detrital deposits on the bank of the Mahānadī in Orissa.

Diamond

Now let us pass on to the most outstanding gem mineral that India ever produced: diamond. C. Ritter in his *Erdkunde von Asien* (Vol. IV, part 2, p. 343, 1836) collected various scattered reports on Indian diamond. Valentine Ball provided a more exhaustive account of diamond works in pre-modern India. (Ref. no. 4, pp. 1-50)

C. Ritter arranged Indian diamond mines known to him in five groups, in order from south to north:

1. The *Cuḍḍapah Group* on the Penner river including ancient mines of Coṇḍāpeṭṭa, Munimaḍāgu, Wajra Kārūr, etc.
2. The *Nāndial Group* between the Penner and the Kr̥ṣṇā river including the mines of Banaganapalli, Rāmulkoṭa, etc.
3. The *Ellore* or *Golkoṇḍā Group* on the Kr̥ṣṇā. This includes famous mines of Kollūr, Partial, Muleli or Malavilly.
4. The *Sambalpur Group* on the Mahānadī river. In this group Tavernier's Sumelpur or Semah/Semul on the North Koel river as well as Wairagarh of the Central Provinces were included.
5. The *Pannā Group* in Bundelkhaṇḍ.

Bauer's map of the ancient diamond fields of India[17] is reproduced in *fig*. 2.1. We would discuss diamond mining in pre-modern India from a chronological point of view and on the basis of literary and archaeological evidences.

Bauer commented[17] that the working of the most important diamond deposits in India known at present 'date back not to very remote periods, probably in all cases subsequent to the year AD 1000 and sometimes much later'. Though we do not have specific C-14 dates, we disagree with Bauer on the basis of literary

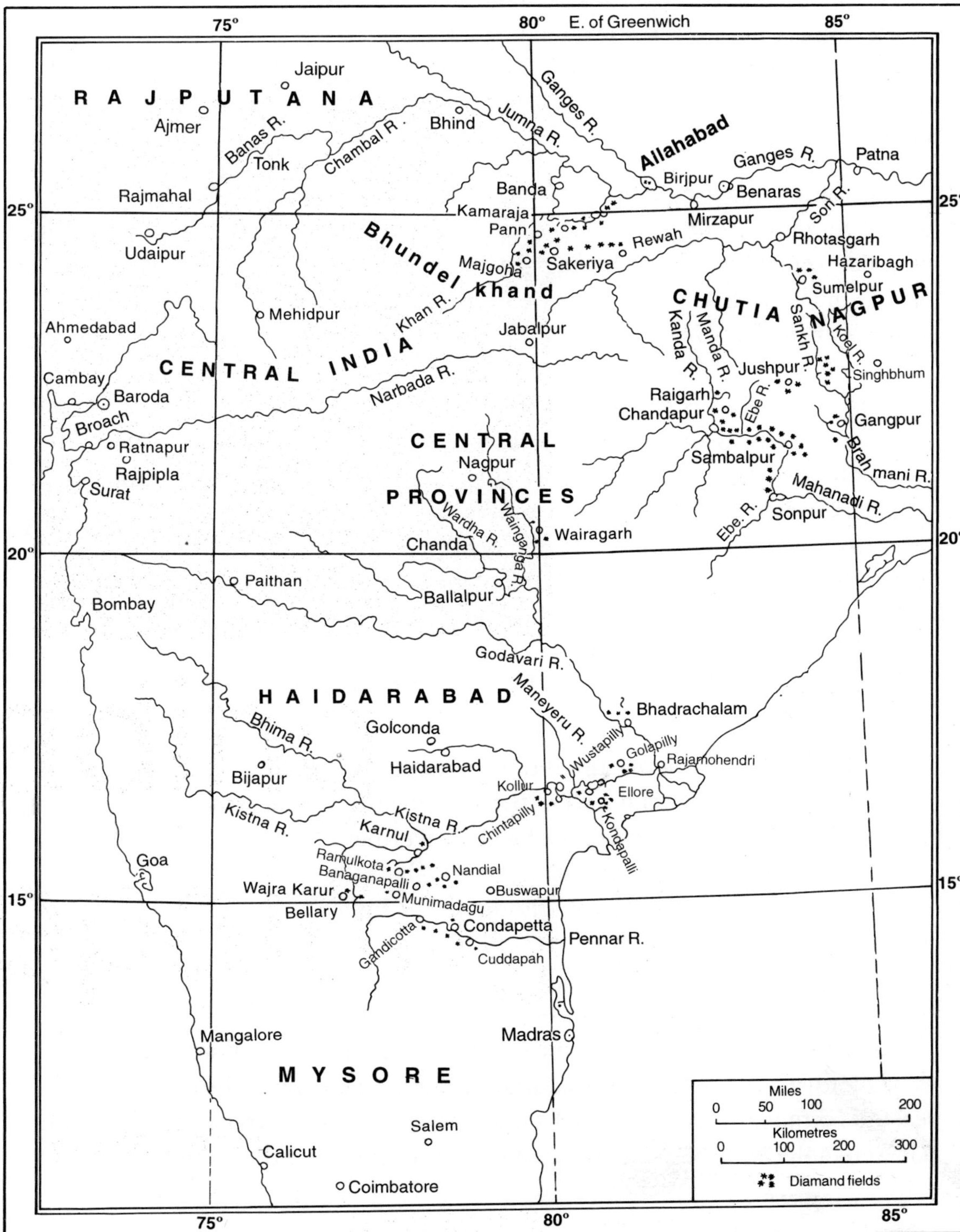

Fig. 2.1: The Diamond-Fields of India (Taken from Bauer[17])

records. The currently known diamond fields in India are shown in *fig*. 2.1.

Kauṭilya's *Arthaśāstra* of the late fourth century BC was probably the first text to describe the Indian diamond or *vajra* and the mode and area of its occurrence (2.11. 37-41). The mine and stream deposits were listed as their sources.

It is hard to identify precise locations of place names mentioned in *Arthaśāstra*, *Bṛhat-saṁhitā*, *Ratnaparīkṣā*, *Agastimata*, etc., regarding diamond fields, but we may conjecture the following locations: Wairagadh (ancient Vajragṛha) eighty miles south-east of Nāgpur on the Sath river, a tributary of Weinganga (Vena Gaṅgā of the *Bṛhat-saṁhitā*); the Kośala region of Ākarāvanti around the famous Pannā in Madhya Pradesh; the region around the Golkoṇḍā mines, formerly known as Mātaṅga; the Pauṇḍa or the Choṭā-Nāgpur area around Soumelpur; the Kaliṅga alluvial resources from the Mahānadī valley, the Sambalpur district, the Koel river Hīrākuṇḍ (literally 'diamond mine'), etc. Diamond-washing has been traditionally done by the tribes of Śavara (Sambalpur area) Kol (Choṭā Nāgpur), Goṇḍ (Madhya Pradesh), etc. Ptolemy had mentioned the country of the Śavaras and Sambalaka (Soumelpur) and of the Maṇḍalai (Muṇḍas) as the sources of Manada (Mahānadī).

ANCIENT DIAMOND WORKS IN THE GODĀVARĪ-KṚṢṆA-PENNER VALLEYS

We have earlier mentioned the accounts of Muslim and European travellers related to Indian diamonds up to the period of the middle of the seventeenth century. During the second half of the seventeenth century, some accounts of Indian diamond mines were published by Tavernier (1676) and Henry Howard (who later attained the honorific title, Earl Marshal), the latter in the *Philosophical Transactions*, Vol, XII, 1677, p. 907 as quoted by Ball (Ref. no. 5, p. 352). Howard enumerated[18] 23 mines in the kingdom of Golkoṇḍā and 15 in the kingdom of Bījāpūr. Some of the important Golkoṇḍā mines reported were Kollūr (somewhat exhausted by 1677), Vajra Kārūr earlier annexed from the Vijayanagara kings, Munimaḍāgu in Karnaul (40 to 50 fathom deep mining), Malavalle, etc. In Bījāpūr area, there were mines of Rammalakoṭa, Banaganapalle (37 mile south-east of Karnaul). Howard, later Earl Marshal, mentioned in his paper that in Golkoṇḍā the miners and merchants were much oppressed, and in a miserable state of poverty; they suffered from tyrannical squeezing and heavy duties on provisions.

During the commencement of the seventeenth-century (around AD 1610), a Portuguese trader had good luck in obtaining a 435 carat diamond around Wajra Kārūr in the Bellāry district (Mir Jumla later took possession of this mine in 1640 from the local kings), and celebrated the occasion by engraving a Telugu poem on a stone tablet praising the lucky Currure.

William Methold reported the discovery (finding by chance) of the Kollūr mine on the Kṛṣṇā river sometime during 1622-6. Some 30,000 to 60,000 persons (township population 1 lakh) were engaged in bailing out the mines by hand, a tedious operation. The wage earned was ridiculously small, one rupee per month, forcing the workers to steal the gems occasionally. The king had farmed out the area to the prospector-merchants, reserving the samples above 10 carats for himself. In 1622, the Moghul emperor demanded a tribute of 3 lbs weight of the finest diamonds when the operation was temporarily closed.

Tavernier described the diamond-processing operations at Kollūr (lat. 16°42'30", long. 80°5') and Rammalakoṭa (lat. 15°34', long. 78°3'15"). In the former site, the processing of the alluvial bed started after ceremonial prayers. The men excavated the earth — 10 to 14 ft in depth — till the water-table was reached. The earth was washed in 2 ft high enclosures with exit openings for water and the slimes. The women and children assisted in washing the clay. The residual coarse material was dried in sun and then winnowed to remove the fine components. The coarse material, thus obtained, was pounded by wooden pestle to liberate hard diamond if any, which were looked for. Tavernier commented:

> Formerly, instead of using wooden pestles for pounding the earth, they pounded it with stones which regrettably produced so many flaws in the diamonds.

The Kollūr mine became exhausted by 1673.

The technique adopted at Rammalakoṭa (and also later in the neighbouring Banaganapalli) was entirely different. The diamond veins in the conglomerate rocks were mined with small irons, crooked at the ends, 'which were thrust into the veins to draw from them the sand or earth and then placed in the vessels for washing'. The directions of the veins were followed to obtain the cleanest and the whitest diamonds. The heavy iron crowbar often fractured the diamonds, giving rise to flaws. 'This is the reason why so many thin stones (*foible*) came from this mine.'

Tavernier also described diamond-cutting on the site by steel-wheel, aided by water, oil and diamond dust. He observed:

> The Indians are unable to give the stones such a lively polish as we give them in Europe: this, I believe, is due to the fact that their wheels do not run so smoothly as ours.

There was brisk business around the mines 'conducted with freedom and fidelity'. Two per cent on all purchases was paid to the king. Even children of age 10 to 16 were proficient diamond-testers handling big and defective specimens and cabuchons (diamond polished but not cut) with equal ease (Ref. no. 5, pp. 43-51). Visitors like Tavernier were courteously treated.

Some 150 years later, Benjamin Heyne visited the neighbouring Banaganapalli mines in 1808. His writing (*Tracts*) was quoted by King:[19]

> The quartzites of the Banaganapalli group form a cap resting on a much older set of shales and traps with some limestone bands. . . . The quartzite covering is from 20 to 30 feet in thickness, and it is pierced here and there by shafts of 15 feet or less, from the bottoms of which nearly horizontal galleries are run to get at the seams of diamond gangue. . . . In the mines, six to eight inches thickness diamond layers were scooped out, and the clayey conglomerate easily broken up. The mass was pounded up, washed, shifted and laid out to dry on prepared floors. The residue of clean sand was carefully examined by the women and children of the working parties, for the precious gems.

The ancient diamond mines in the Bhīma-Tuṅgabhadrā-Kr̥ṣṇā-Godāvarī valleys in the Andhra Pradesh region (*vide fig.* 2.1) have been specially studied and reported by Voysey[20] (1833), King[19] (1872), Munn[21] (1929), Dutt[22] (1953) and Rao[23] (1969). Dutt has drawn attention[22] to the occurrence of diamond mines in the Andhra Pradesh region in three forms: (1) in river gravels, (2) sedimentary rocks or detritals, and (3) in the Archaean crystallites. Except in river alluvials, no occurrence of diamond has been reported from the eastern portion of the sedimentary area. On this basis, Dutt has postulated[22] that the west of the Cuḍḍapah-Kurnūl rocks or the Archaean crystallites, which produced largest diamonds, was the probable geological home of the regional diamond, and from this area the gem was transported by the rivers winding easterly. Rao has reported[23] a large number of ancient diamond mines in a narrow area of the Kr̥ṣṇā valley (between latitude 16°02' to 16°39' and longitude 78°19' to 80°25'), those of Kolhāpur, Bollaram, Amargiri, Somsil, Maddimaḍāgu, Thigamanpenṭa, Partial, etc.

Some of the ancient works were reported by Howard.[18] A common feature of many of these works was the use of stone cisterns, which were washing places (as described by Tavernier in connection with the Kollūr mines). The material excavated from the pits used to be soaked in water in these cisterns and washed several times by letting out the water through the holes at the bottom of the walls. The residue was then examined for diamonds after drying, and in some places, after gently powdering and winnowing the same.

The alluvial workings around the Penner river in the Cuḍḍapah district are also worth mentioning. Ball referred to the old workings of Kanuparti (14°34'; 78°49'); the Coṇḍapeṭṭa of Dr. Heyne and Captain Newbold; Obalampalie (14°33'; 78°51'35") and Dorlampalle on the map. These two sites are near Chennur, but on the opposite banks of the Penner river. Captain Newbold wrote (Ref. no. 4, pp. 9-11):

> At Coṇḍapeṭṭa the mines are generally of a square form, and 4 to 12 feet deep. The stratum cut through is of cotton soil, mixed with small grains of quartz, generally from 3 to 10 feet thick, which rests immediately on a bed of rolled stones of various sizes, in which diamonds are found generally loose, but sometimes adherent. . . . The process of mining consists merely in digging out the rolled pebbles and gravel, and carrying them to small square reservoirs raised on mounds having their bottoms paved with stones, and washing them carefully. . . . The diamonds are easily recognised in the moist state by their peculiar lustre.

Newbold's description referred to the workings which were administered by the East India Company but they followed ancient methods:

> In former days all the diamonds produced were carried for sale to Golkoṇḍā. In those times very large diamonds were found, but subsequent to the British ascendancy — which according to the superstitious natives, is by no means pleasing to the tutelary deities of the mines — few of any value have been found, probably in consequence of their being less looked after. However, lately, in 1839, a fine diamond of the kṣatriya or roseate caste was dug from the Obalampalle mine (discovered around 1755 AD according to Dr. Heyne), exceeding a gold *pagoḍā* (16 carats) in weight, which was sold for Rs. 1450.

Both Dr. Heyne and Captain Newbold mentioned, while describing their visits to these mines, that 'the local miners objected to their approaching them on horseback, as it would, they said, irritate Ammawarū or Lakṣmī, the sanguinary goddess of riches, who was the patroness of the mines'. Newbold witnessed sacrifices to propitiate her and noted the Telugu terms for different kinds of stones which the miners referred to as associates of diamond: *binga bendu*, transparent quartz; *patcha bendu*, epidote; *baggira*, jasper; *karla*, basalt; *yerra bendu*, sandstone; *korund* or corundum, which was considered to be the best sign, indicating possible association with diamond. A brāhmaṇa class diamond fetched a price almost the double of śūdra class of equal weight.

The valleys of Tuṅgabhadrā, Bhīma, Kṛṣṇā, Godāvarī and Penner had provided enough Archaean crystalline diamonds (*vide fig.* 2.1) to the south Indians for centuries. This has been reflected in many writings, starting from Amir Khasrau's description of the sack of Wāraṅgal and the loot of diamonds (early fourteenth century) and ending up to Heyne and Newbold's narrative of the superstitious but crafty Telugu-speaking miners on the Penner river (nineteenth century). Ball noted (Ref. no. 5, pp. 349-51) ancient diamond mines in 52 specific localities in this region (latitudes 14-17° and longitudes 77-80°).

THE ANCIENT DIAMOND MINES OF BENGAL-BIHAR-ORISSA

In his writings on the 'Diamond mines of (greater) Bengal', Valentine Ball (Ref. no. 4, pp. 24-37; Ref. no. 5, pp. 354-9, Appendix III) drew attention to the three distinct localities in the Bengal-Bihar-Orissa region which produced diamond in the pre-modern period: (1) the Sema on the Koel flowing northwards, a tributary of the Son; (2) the Sankh river flowing southwards on the other side of the watershed, one of the tributaries of the Brāhmaṇī; and (3) Sambalpur-Hīrākuṇḍ area on the Mahānadī further south (*vide fig.* 2.1).

About 81 miles south of Rohtāsgarh on the bank of the Koel, we have the remains of an old town Semah (lat. 23°35', long. 84°21') which was Tavernier's Soumelpour, a corruption of Śimulpur (the *śimul* is the silk cotton tree, *Bombax malabaricum*). On the other side of the watershed from which Koel originates to flow northwards, the Sankh river flows southwards to join Brāhmaṇī. Since both the rivers produced alluvial deposits containing diamond, the origin of the gem-mineral must have been the common watershed or the bedrock of the hills (*vide fig.* 2.1).

This area was probably known to Ptolemy as Sambalaka of Maṇḍalai (Muṇḍās, who also produced Muṇḍaloha or iron around Lohardugga) and later to the Moghuls, as Kokrah an abbreviation of Kokko-nāge or Choṭā Nāgpur.

Blochmann reported (*Journal of Asiatic Society of Bengal*, Vol. XL, p. 113) two notices of Kokrah — one in *Akbarnāmāh* and the other in *Tuzuk-ī-Jehāngīrī*. In 1585, Madhu Singh ruler of Kokrah accepted subservience to Akbar. In 1616, Ibrāhim Khān, Jehāṅgīr's governor at Bihar, overran Kokrah and took possession of its diamond resources. *Jehāṅgīrī* records:

> This district belongs to *Sūbāh* Bihar, and the river which flows through it yields diamonds. When the river contains little water, tumuli and hollows are formed. The diamond-diggers know from experience that those tumuli chiefly, contain diamonds over which insects, called by the Hindus *jhinga*, hover. They pile up stones on all sides of the tumuli, and then cut into them with hatchets and chisels and collect the diamonds from among the sand and stones. Sometimes diamonds are found of the value of a lakh of rupees each. When I (Jehāngīr) appointed Ibrāhim Khan governor of Bihar, I told him to invade the district and deprive the *zamīndār* Durjan Śal of all the diamonds in his possession.
>
> The district is now subject to me. All diamonds found in the river are forwarded to courts. Only a few days ago, a diamond arrived which had a value of Rs. 50,000. There is one which look like sapphire: I have never

> seen a diamond of such a colour. My lapidaries fix its value at Rs. 3,000, though they would give Rs. 20,000 for it if it were quite white and stood the full test.
>
> — Quoted by Ball Ref. no. 4, pp. 25-6

Blochmann believed that the diamond-river alluded to, was the Sunk. To the present day a spot in the Sunk is pointed out by the inhabitants as being a place where diamonds were washed for. The present author feels that Jehāṅgīr's diamond-river could be as well (north) Koel, corresponding to Tavernier's Soumelpour or modern Semah. Tavernier wrote:

> Soumelpour is a great town, the houses whereof are built of earth and covered only with branches of coconut trees. The Raja lives half a *coste* (little over a mile) from the town in tents, set upon a fair rising ground at the foot whereof runs the Koel.
>
> This is the manner in which they search for diamonds. After the great rains are over (they wait till the end of January) there pour forth 8000 persons — men, women and children — those who are able to work. They that are skilful know by the sand whether there be any diamonds or not.
>
> Where they believe there are diamonds, they encompass the place with stakes, faggots, and earth; they go about to make the arch of a bridge to drain all the water out of that place. They dig out all the sand for 2 feet deep, and spread upon the side of the river in an enclosure of about 1½ ft high. Then they wash it, and shift it, doing other things as they do at the mines.

The Rājās of this place had rebelled against the Moghuls from time to time. Ball refers (Ref. no. 4, p. 29) to a large picture by a contemporary artist, representing the attack on the fort of Palamow in 1660 by Daud Khān and depicting a figure of the *Zamīndār-ī-kan-ī-almās*, or lord of the diamond mine, who is considered to have been a Kol Nāgbanśī Rājā.

SAMBALPUR DISTRICT DIAMONDS

Tavernier's Soumelpour or Semāh is not to be confused with Sambalpur in Orissa on the Mahānadī. Ptolemy's Adamas or diamond river in the land of the Śavaras

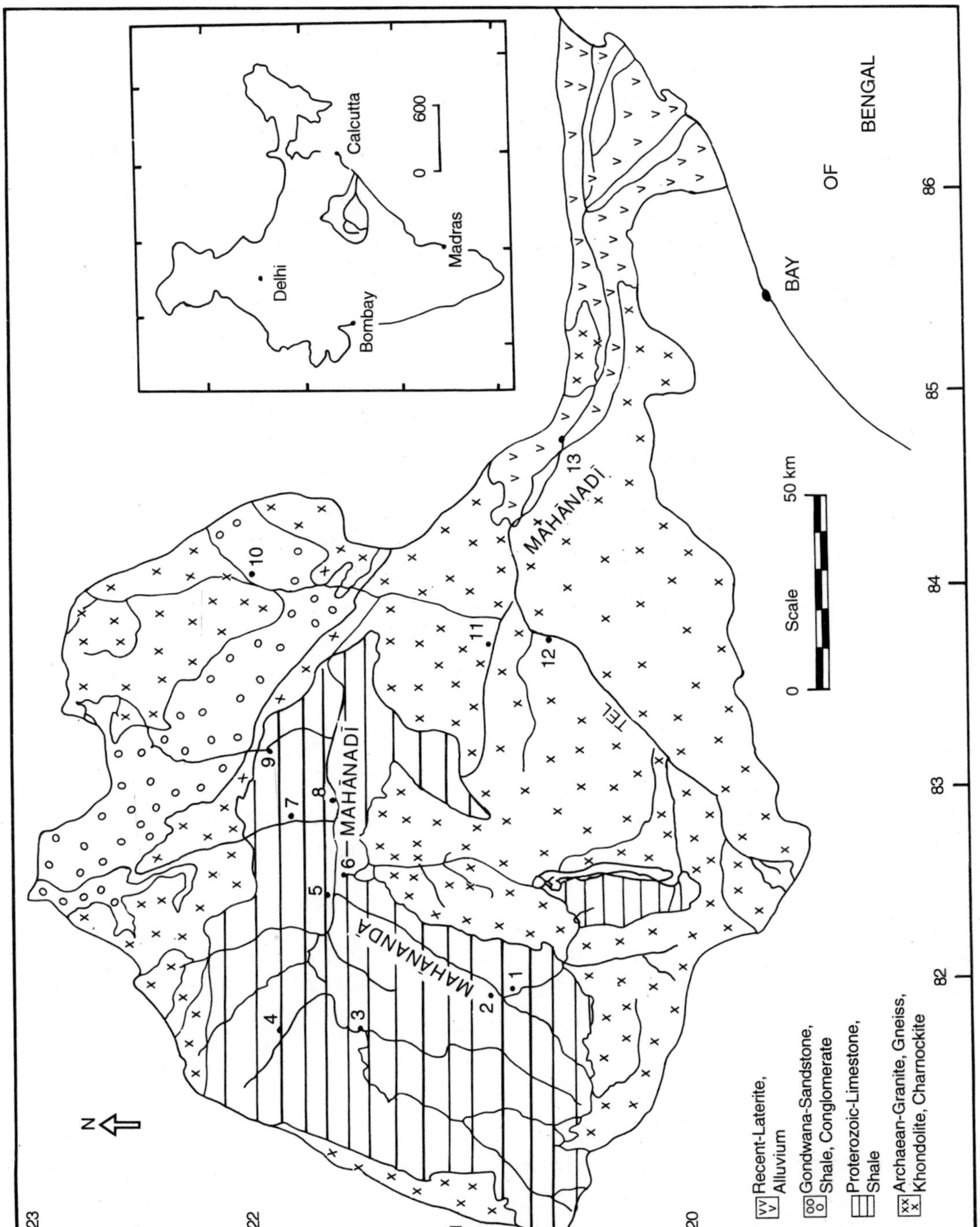

Fig. 2.2: Geological Map of the Mahanadi Basin (Taken From Panigraphy[25])

Locations: 1. Baronda on Pairi (Stream) 2. Rajim on Mahānandā 3. Simga on Seonath 4. Andhiyarkore 5. Jondhra 6. Rampur 7. Bamnidih 8. Basantpur on Mahānadī 9. Kurubhata on Mand 10. Sundergarh on Ib 11. Salebhata 12. Kantamal 13. Tikarpara on Mahānadī is most probably the Mahānadī. The tributaries such as the Mand and then the Ebe or Ib join the Mahānadī flowing from the west to the east (*vide fig.*2.2). Diamonds have been found not on the river Ebe but on the Mahānadī riverbed, west up to Cāndapur. Ball conjectured (Ref. no. 4, pp. 34-5) that the quartzites, limestones, the sandstones, and the shale of the Barāpahār hills provide the laterite or red earth matrix in which diamonds are found. Along with diamond, many other naturally occurring precious stones are washed down the Mahānadī river tract.[24] Ball recorded that in addition to diamonds, pebbles of beryl, topaz, carbuncle, amethyst, carnelian, and clear quartz used to be collected in the Mahānadī — this is still being done as reported by Ghosh[24] — but it is probable that the matrix of most of these exists in the metamorphic rocks on which Hīrākuṇḍ is located, and is, therefore, distinct from that of the diamonds (Ref. no. 4, p. 36).

The first visit to the Sambalpur diamond working ever to be reported was undertaken by Mr. Motte in 1766 at the behest of Lord Clive (Ball quoted *Asiatic Annual Register*, London 1799 and Hamilton's *Hindustan*, Vol. II, p. 20). A servant of the Rājā at Sambalpur showed Mr. Motte how diamond was being recovered from the red earth (apparently by washing). The digging operation in the mountains had been suspended on account of the Marāṭhā raids.

Dr. Voysey (1823) and P. Breton (1825) visited the same topography going further west up to the confluence of Maṇḍ and the sites of Rāigarh, Jaśapur, etc. Breton reported that in 1809 a large (over 200 carats) vaiśya quality diamond was recovered at Hīrākuṇḍ, but this was confiscated by the raiding Marāṭhās. In 1836, Karl Ritter published his careful compilation on Indian diamond and other gems (*Erdkunde Asien*, Vol. IV, part 2, p. 343).

In 1840, Major J.R. Ouseley published his note on the process of washing for the gold dust and diamonds at Hīrākuṇḍ, which literally means diamond-mine (*The Journal of the Asiatic Society*, Vol. VIII, January-December 1839, published in 1840, pp. 1057-8). Hīrākuṇḍ is located about seven miles from the eastern end of Sambalpur and four miles downsteam from the confluence of the Ebe and the Mahānadī. Gold and diamond are available on the Mahānadī down up to Sonpur.

Ball elaborated Ouseley's description with more details which he collected from the oldest of the *jhiras*, or washers at the village of Jhunan, near Hīrākuṇḍ, who described to him the operations during Rājā's time that is before 1850 when the British took over Sambalpur. Now we quote Ball (Ref. no 4, pp. 35-6) and Ouseley.

In the centre of the Mahānadī, the island of Hīrākuṇḍ is about 4 miles long, and for that distance separates the waters of the river into two channels. In each year, about the beginning of March, when the level of water was approaching the lowest, nearly five thousand people assembled and raised an embankment across the mouth of the northern channel, its share of water thus being deflected into the southern. In the stagnant pools left in the former, sufficient water remained to enable the washers to wash the gravel, accumulated between the rocks.

The men and the children brought the gravel and sand in wooden trays and placed it in the trough, which is open at one end, with a gentle inclination towards the river. The women sat on the edge of the river. With their left hands they stir up the gravel, and with the right, pour water out of a wooden basket-looking bucket gently over the upper end; it runs out in the river, the larger pebbles and gravel are thrown over. The finer sand, on the trough being full, is rewashed, removed into wooden trays, and by dipping them under water, and shaking them about, the gravels gradually fall over, leaving only gold dust.

> They detect the diamonds (in the finer gravels) at a glance, as they wash. Once I (Ouseley) saw about the size of a large grain of wheat, clear and bright; but these are the Raja's property. The gold they are allowed to dispose of at 12 to 15 rupees per *tolā*.

Ouseley 'fancied' that recovery of gold and diamond could be done by 'mechanical principles which would, by reducing the manual labour, make the speculation highly profitable'. He recommended a vibrating graduated wire-sieve (6 ft long. 2 ft wide) washing machine, larger at the top, and smaller as the sieves approached the bottom, below which a wooden tray could be kept. "If the machine were placed in a gentle stream, the gold dust would be found in the tray." The diamond samples could be looked for in the finer gravel fraction. Ball suggested that the southern channel of the Mahānadī, which was never emptied in the Rājā's time, could yield a larger number of diamonds than the northern.

> The stronger current in it would be more efficient in removing the substances of less specific gravity than diamonds, while the rocks and deep holes in it afford admirable means for the retention of the latter. It would be more difficult to divert the body of water greater than that which flows in the northern channel; but the result in a greater harvest of diamonds would probably far more than compensate for the greater expenditure incurred. — Ref. no. 4, p. 35

These suggestions were never tried in the nineteenth century, but in the context of the Hīrākuṇḍ dam and other modern developments, we may try them as we approach the twenty-first century.

As a matter of fact, the banks of the Mahānadī in southern Orissa has been found to be a virtual treasure trove of precious and semi-precious stones — not only diamond but also many other gems.[24] Panigrahi has reported[25] that the physical erosion rate in the Mahānadī basin (*vide fig*. 2.2) varies between 146 to 707 tons/sq km/year, the uppermost value corresponding to the upland tributary Pairi ending up in Mahānadī. The ratio of physical to chemical erosion is very high varying between 5.2 to 43.0. The huge mass of solid eroded from the archaean granite and sandstone/limestone shale deserves to be monitored and beneficiated for a total recovery of precious stones.

The diamonds washed down from the Madhya Pradesh Kimberlite rocks are still found in crevices in the rocks at Boudh. Early 1992, a 57-carat blue diamond has been found in the Mahānadī, reminding one of Jehāṅgīr's comment on a Sankh river diamond 'which looked like sapphire'. The recent findings of : aquamarine in Badmal and Sangamora in Bolaṅgīr; rubies and iolite in Jilingdhar and Hiñjlibahal in Kālāhāṇḍī; chrysoberyl in Korapūṭ; sapphire in Amherā; garnet, rare viloet rhodolite at Siali and many others have made the Orissa Government arrange for high resolution aero-magnetic survey on ground water as well as mother lodes for gems in this area and seek assistance from the United Nations Development Programme in setting up semi-mechanised panning and screening facilities and mobile washing plants on the banks of Mahānadī.[24] This would fulfil the dreams of Major Ouseley and Valentine Ball!

Diamond in Central India

The ancient mine at Wairagarh (lat. 20°26', long. 80°10') situated about 80 miles to the south-east of Nāgpur is of considerable antiquity. The diamonds have been found in the yellowish earth lateritic grit with quartzose and metamorphic rocks in the vicinity.[26] *Kauṭilīya Arthaśāstra*'s (2.11.37), *Sabhārāṣṭra* and Varāhamihira's *Bṛhat-saṁhitā* (80.6) *Veṇātaṭe* probably refer to diamond mines in this area. Veṇā is modern Weinganga or Vāṇagaṅgā. Wairagarh mine is on the Sath river which ultimately joins Weinganga. Ajay Mitra Shastri has quoted[27] several Khāravela and Cola inscriptions to suggest that Wairagarh was known in the ancient times as Vajrākara or Vajragṛha, the mine or abode of diamond.

Most probably the mine, taken possession of by Ahmed Shāh Walli Bahamanī in 1425 (as recorded by Ferishta), was this mine at Wairagarh. *Āin-ī-Akbarī* mentions this place as Beiragarh (Gladwin's translation, Vol. II).

R. Jenkins reported in 1827 that this mine was not yielding sufficient returns. During Rāghojī Bhonsle's time the mines in the neighbourhood were worked at considerable expense, but since only a few diamonds were found, they were finally given up. Ball suggested that the great basin of Lower Vindhyan or Karnul rocks, which occupies the upper portion of the Mahānadī valley in Madhya Pradesh, stretches into the neighbourhood of Wairagarh (*vide fig.* 2.1) and thus the diamond beds of the two regions have common geological genesis.

THE PANNĀ MINES

The mines of Bundelkhaṇḍ are generally known as the Pannā mines, but it is necessary to distinguish the mines near the town of Pannā from those which are situated in other parts of the province. Ball mentioned (Ref. no. 4, pp. 39-49) several amongst a large number of ancient mines: Kāmarīya, Pannā (or Punnāh), Majgoha, Brijpur, Udesna, Sakeriya, Baghin, Myra, Etwa, Bargari, Saya Luchmanpur, etc. — of which the first three were more important than the others.

In this area, the origin of diamond is in volcanic pipes. Ultra-basic diamondiferrous plug is made up of a highly serpentised rock akin to the kimberlite of South Africa. The plug consists of greenish clay for the first few feet of depth. Deeper it becomes harder and is found to be greenish 'tuff' (volcanic rock). Crystals of quartz, calcite and olivine are bedded in a serpentinised groundmass.[28] The green mud is actually quartz containing hydrated nickel silicate, known as chrysoprase, prase, plasma and also 'mother of emerald'.

Kāmarīya mine was visited and described by Capt. Franklin (1827) and H.B. Medlicott (1860). The diamond stratum, locally called *kakru*, consists of a conglomerate sandstone made up of pebbles, 1/8 to ½" in diameter, embedded in a rather fine matrix. There were numerous old pits in this area and around Bungla and Bābupur, 10-20 ft deep. There was an old pit south-west of Kāmarīya which was 30-40 ft deep and abandoned, because the bottom was filled with water. Franklin considered the green pebbles to be an indicator of the presence of diamond in the bed.

Pannā: The mines, situated about 2-3 miles to the north-east of Pannā (lat. 24° 43'30", long. 80° 15') were frequently visited and reported during the nineteenth century: Dr. F. Hamilton (1813), Capt. Pogson writing *The History of Boondelas* (1826), Capt. Franklin (1827), J. Adam (1840), H.B. Medlicott (1860), M. Rousselet (1874), W.L. Wilson (1877), — all quoted by Valentine Ball in 1881.

Ball described:

> In the Panna mines, although the diamond seam is deeper than elsewhere, it is not reached by a shaft, but the miners go to the immense labour of excavating great pits, 25 ft in diameter, and often over 30 ft deep, for the sake of the small patch of diamond conglomerate.
>
> The almost naked miners enter the pit by an inclined plane and work knee-deep in water. The stones and mud which they excavate are put into small baskets which are drawn up by hand. The Persian wheel turned by four bullocks is insufficient in power to thoroughly drain the place; water was lifted by a band of earthen vessels. — Ref. no. 4, p. 40

The above-described system of mining in practice at Pannā was illustrated through a sketch drawn by Jules Schaumburg, who accompanied M. Rousselet to Pannā in the year 1867. The illustration was published in Rousselet's *L'Inde des Rajahs* (1874, pp. 441-3) and reproduced by V. Ball (Ref. No. 4, facing p. 40).

(Vide Plate 2, p. 198)

Ball continued:

> The stuff on reaching the surface is placed on stone slabs and is searched, the searchers being under the charge of guards, as are also the miners.

Majgoha, now known as Majhgawan, is the westernmost point in the diamond area and was visited by Franklin and Medlicott. The former described the mine to be a huge basin, like an inverted cone, 100 yards wide and about 100 ft deep. The diamonds occurred with the green mud, deeper below a depth of 50 ft. The nineteenth-century appliances did not permit large scale mining.

A survey around 1960 suggested that the total reserves of diamondiferous 'tuff' material at Majhgawan would be about 56 million tons with a diamond content of about three million carats; this could be recovered by a combination of diverse and modern mineral beneficiation techniques. As a matter of fact centralised diamond-dressing plants were recommended to treat diverse diamond-bearing ores in the country.[28]

The recent explorations thoroughly justify the observation made by Franklin one and a half century ago that 'an improved system of mining (at Majgoha) might be applied with good effect'. We are indebted to him and the other nineteenth-century British geologists and travellers for meticulously noting the ancient diamond-mining practices in India.

Valentine Ball quoted an anonymous report on Pannā diamonds published in an Indian newspaper and provided further details. Whereas in the time of Akbar the value of the diamond mines was about Rs. 8 lakh a year, in 1750 the profit came down to Rs. 4 lakh; and Franklin reported in 1827 the annual produce to be worth Rs. 1.2 lakh only. A very insignificant amount was paid in wages to the miners. The monthly expenditure incurred in one mine was Rs. 122 only, 20 builders and 15 waterwomen getting Rs. 2 per person per month, 4 sepoys each at Rs. 3 and the rest (Rs. 40) spent as digging cost.

The poor miners working at the *ghaira* (deep mines) and *chila* (surface pits) were inhumanly supervised by *tuadar*s (masters of the mines) who in turn were mercilessly exploited by the Rājā through oppressive rules. To counter this chain of oppression, the *tuadar*s would often smuggle valuable gems to the loan-giving bankers or *mahājans* who would carry on clandestine trade of diamonds between Mirzāpur, Vārāṇasī, Allāhābād and Jabalpur. The miners would try their best to pilfer and conceal the best gems, and even 'in case of emergency, swallow them'!

The decline in the diamond trade

Valentine Ball asserts (Ref. no. 4, p. 48) that there was no real exhaustion of the localities where diamond mining was possible. On the contrary, the diamond beds were extended far more than the ancient miners ever supposed. They acted on chance discoveries and rule of thumb knowledge. That they systematised their observation about the diamond-bearing strata and applied them to distant tracts is extremely doubtful. There is no evidence that any intercourse or communication took place between the workers at distant localities. Ball hoped that scientific guidance would improve diamond production in India.

Dutt attributed[22] various reasons to the decline of Indian diamond industry: exhaustion of diamond-bearing rocks, water trouble in the excavations, oppressive nature of the mining and political administration, absence of systematic prospecting operations, superstitions amongst the workers and the discovery of diamond fields in other parts of the world. India lost its monopoly in diamond trade in 1728 when Brazilian mines were first exploited. In 1870 the South African mines monopolised the global market in this precious gem.

Concluding remarks

The pre-modern Indian traditions in gem production were doomed to extinction in view of the inefficient technology and management which also hovered over the non-gem mineral and metallurgical industries as well. Thus the comments made by Ball (Ref. no. 4, p. 48) and Dutt[22] on ancient diamond mining in India have general applicability. The common issues may be discussed later in a broader framework.

Our present discussion on gem minerals in pre-modern India indicates a remarkable continuity in traditions from the sixth to the end of the eighteenth century. This has been adequately testified by the Arab, Persian and the European travellers. The British travellers and scientists of the eighteenth and nineteenth century witnessed pre-modern technologies on mining and processing of gem minerals and left detailed accounts which have been brilliantly summed up by Valentine Ball.[4]

The Sanskrit words for gem minerals, culled from the *Ratnaśāstra* and the *Rasaśātra* texts, were compiled and tabulated by the present author.[1] Some of the Arabic/Persian words used in medieval India and their approximate or likely English equivalents[15,29,30] are provided in Table 2.1.

Mining and processing of gem minerals used to be done for the affluent part of the Indian society and the outside world. For low quantum of output, a very large number of poor people worked under appalling conditions. Technological levels were primitive and almost no attempt was made to upgrade them. Interactions between workers of different trade guilds, and between workers and intellectuals were negligible. The scholars of the *Ratnaśāstra*s and the Muslim gemmological texts hardly advanced their earlier knowledge on the subject. In such an apathetic atmosphere, the ill-paid and ill-fed Indian workers toiled in the mines and produced exquistite art-jewelleries for the whole world to marvel at.

Table 2.1

Some Persian / Arabic / Hindustānī terms regarding gem minerals used in Pre-Modern India

(Taken from Holland,[15] Prinsep[29] and Tagore[30])

Abu-Is'Haqi	Father of Isaac or 'genuine' turquoise, product of Nishapur in Khorasan, discoloured by heating. Probably this is *calaite* variety of turquoise containing 18 per cent water.
Akīk	agate (also carnelian in Yemen)
Akīkīya	gem cutter and polisher
A'lmāsh	also *Almās* — diamond. See *Māsh*
Āyn-ul-Hireh	cat's eye, chatoyant sapphire
Badakshānī	turquoise different from Abu-Is'haqi or Nishapuri variety. Withstands heat or fire without alteration. Could be odontolite or bone turquoise. Badakshānī ruby is Balas-ruby, Kauṭilya's Bālasūryaka.

Basud	coral-like porous material devoid of gelatinous mass, available in Yemen, Maldive sea-water.
Bawa Ghori	onyx
Begri	lapidary
Bellāor/Bellor / Bilhaur	see *Bilor*
Berūj	beryl from *vaidūrya* (Sanskrit)
Bidhiyā	who bores holes through precious stones
Bilor	rock crystal
Cheshamdar	quartz cat's eye, see also *Dola*
Chuṇi	ruby weighing less than one *rattī*
Dola	quartz cat's eye
Dschemst	amethyst
Ebrendshe	gold marcasite
Ferozā / Firozeh / Firuse / Pēroj	turquoise. See Abu-Is'Haqi and Badakshānī
Gulābī	rose-coloured corundum from Rewah
Hakīk	agate. See *Akīk*
Hukkāk	cutters and polishers of gem
Huzur	gem (Arabic)
Ilmās	actually *A'lmāsh* or diamond
Jaschep	jasper
Jauhar/ Jawāhir / Jawahar	jewel
Juzā	cat's eye found in the mines of *Akīk* or agate
Kamānī	a bow used for working a gimlet in gem-stone boring (Hindustānī)
Kash	from *Yu she*, Chinese for jade which comes from Eastern

	Turkistan. Hence the name of the place is Kashgarh
Korund / Kūrūn	corundum from Kuruvinda (Sanskrit)
Lājaburud/Lājaward	blue lapis lazuli. In Sanskrit *rājāvarta*
Lāl	red-coloured balas or spinel ruby
Lāl-Badakshānī	rose-coloured ruby of Badakshān, capital Balkh. Balas-ruby, *Bālasūryaka* in *Arthaśāstra*
Lāl-mīnā	violet-coloured almandine ruby
Lāl-piāzī	onion-coloured, reddish yellow ruby
Lāl-pun nīla	sapphire with a tinge of red
Lāl-rūmānī	scarlet or pomegranate-coloured ruby, usually from Myanmar
Lulu	pearl (Arabic)
Mānak / Mānik	ruby weighing over one *rattī*. See *Chuni*
Mark-ī-Shīsā	marcasite or iron pyrite
Marowarit / Merwarid	*margarita* or pearl
Mash	diamond (Arabic)
Mora	onyx
Murjān	coral, best variety *mūṅgā*
Nīlā	sapphire (blue)
Pitoniā	helioptrope or bloodstone which is green gem (plasma) interspersed with dot of red
Pokhrāj / Pukhrāj	topaz
Rag-ī-lāl	vein of rubies
sane	corundum wheel or stone-cutting wheel, also for polishing gems and sharpening tools
Saṅgtarāśa	stone-cutter

Sard	sardonyx
Senbade	corundum
Sitarei semin	star of the earth or talc
Sonelā	an inferior variety of topaz (golden)
Suabhaji	moss agate
Subj-pun nīla	sapphire with a tinge of green
Sung	gemstone (Persian)
Tāblā	cut rubies (Burmese)
Tanjīn	polished rubies (Burmese)
Torā	yellowish beryl
Turmali	'a kind of *yāqūt*', possibly tourmaline, could also represent zircon, hyacinth, etc.
Upul	opal (many coloured)
Yākūt / yāqūt	hard gem minerals like ruby, sapphire, etc., a generic name.
Yākūt-ahmar	oriental ruby
	arghwānī — hyacinth; *khamrī* — red-wine coloured, amethyst; *Khylī* — asafoetida-coloured; *lahmī* — flesh coloured; *mihrmātī* — striped; rūī — brass coloured
Yākūt-arzaq or Qabūd	sapphire, *nīlā*; *taüsī* — peacock-tail; *nīlī* — indigo
Yākūt-asfar or *zard*	*topaz*, *pokhrāj*, *narinjī* — orange; *turanjī* — citron-coloured
Yākūt-rumānī	bright red ruby
Yākūt-surkh	amethyst or topaz (Gladwin)
Zabarjad/Zuburzud	beryl, green aquamarine; also confused with yellow chryso-beryl and topaz
Zaffre	cobalt sulphide used for painting of glass and porcelain

Zahābī	gold-coloured or yellow beryl
Zamarūd/ Zumārrūd	emerald, green crystalline.

References

1. A.K. Biswas, *Minerals and Metals in Ancient India* in Two Volumes, report to the Indian National Science Academy, New Delhi, 1991, Vol. I on archaeological evidences. Discussions on gem-minerals in most of the 21 chapters particularly in chapter 16 on Indo-Roman trade. Vol. II on literary evidences has chapter 7 on *Ratnaśāstra* (gemmology). Published by D.K. Printworld (P) Ltd., New Delhi, 1996.
2. A.K. Biswas, *Gems and Minerals in Ancient India*, included in the INSA publication History of Technology in India — Ancient Period, Vol. I, 1997, ed. by A.K. Bag, pp. 225-55.
3. Valentine Ball, A Geologist's Contribution to the History of Ancient India, Presidential Address to the Royal Geological Society of Ireland, *Journal of the Royal Geological Society of Ireland*, Vol. V, Part III (New series), 1879-80, pp. 215-63.
4. Valentine Ball, *A Manual of the Geology of India, Part III, Economic Geology*, Geological Survey of India, Calcutta, 1881. 16 chapters, 7 appendices, 11 plates and 6 maps. A superb collection-gem.
5. Valentine Ball, *Travels in India* by Jean-Baptiste Tavernier — edited translation of the original French edition of 1676 in 2 volumes, 1889, Reprint by Oriental Books Reprint Corporation, New Delhi.
6. S.M.R. Ansari, On the Physical Researches of Al-Biruni, *Indian Journal of History of Science*, Vol. 10, No. 2, November 1975, pp. 198-217.
7. Muhammad Habib, *The Campaigns of Alauddin Khilji*: *Khazainul Futuh or Treasures of Victory* — of Hazrat Amir Khusrau — translated and ed. by Muhammad Habib, D.B. Taraporevala and Sons, Bombay, 1937.
8. S.R. Sarma, Ṭhakkura Pherū's *Rayaṇaparikkhā*, Viveka Publications, Aligarh, 1984.
9. Eilhard Wiedemann, *Aufsaetze Zur Arabischen Wissenschaftsgeschichte*, Hildesheim, New York, Vol. I, 1970, pp. 829-80 on mineralogy of Islam, specially p. 867.
10. Tapan Ray Chaudhuri, and Irfan Habib, *The Cambridge Economic History of India*, Vol. I. *c.* 1200-1750, Cambridge University Press, 1982, Reprint by Orient Longman Limited, New Delhi, 1993 (a) pp. 102-24; (b) p. 314
11. M.L. Nigam, Central Asian Contribution to the Art of Jade Carving in India *in* Indo-Soviet Seminar on Scientific and Technological Exchanges between India and Soviet Central Asia in Medieval Period, ed. by B.V. Subbarayappa, Indian National Science Academy, New Delhi, 1985.
12. H. Blochmann, The *Āin-i-Akbarī* of Abul Fazl (edited translation, revised by D.C. Phillott), Oriental Books Reprint Corporation, New Delhi, Volume I, 1977, pp. 15-16.
13. W.E. Begley, and Z.A. Desai, The *Shah Jahan Nama* of Inayat Khan, ed. and tr. Oxford University Press, New Delhi, 1990.

14. A.K. Biswas, Vaidūrya, Marakata and Other Beryl Family Gem Minerals: Etymology and Traditions in Ancient India, *Indian Journal of History of Science*, Vol. 29, No. 2, 1994, pp. 139-54.
15. T.H. Holland, *A Manual of the Geology of India-Economic Geology* by Late V. Ball, Second Revised Edition 1898. Part I on Corundum, pp. 52-67, specially pp. 60-1, and the annexure on the Persian/Arabic terms for gems, pp. 66-7.
16. F.R. Mallett, *A Manual of the Geology of India, Part IV Mineralogy*, Geological Survey of India, 1887, pp. 91-2.
17. Max Bauer, *Precious Stones*, 2 volumes, Dover Publications, New York, p. 141 and p. 144, 1968.
18. Henry Howard, (Earl Marshall), A Description of the Diamond Mines as it was Presented . . . to the Royal Society, *Phil. Trans. Royal Society, 12*, 907-17, 1677 reprinted in *Journal Hyderabad Geological Survey*, *1*, 1, 1929.
19. W. King, On the Kadapah and Karnul Formations in the Madras Presidency, *Memoirs Geological Survey of India*, *8*, 1, 1872.
20. H.H. Voysey, Report of the Geology of Hyderabad (in two parts), *Journal Asiatic Society of Bengal*, Volume 2, 1833.
21. Leonard Munn, The Golkoṇḍā Diamond Mines, *The Journal Hyderabad Geological Society 1*, 1, 1929, pp. 21-62.
22. N.V.B.S. Dutt, Ancient Diamond Mining, in Andhra and its Future, *Indian Minerals*, published by Geological Survey of India, Vol. 7, no. 1, January 1953, pp. 138-50.
23. P.V. Rao, Old Diamond Workings in the Erstwhile Hyderabad State, *Indian Minerals*, published by Geological Survey of India, Vol. *23*, no. 1, January-March, 1969, pp. 1-15.
24. Subarna Ghosh, Precious Hunting Grounds — Mahānadī, a virtual treasure trove of precious Stones, *The Statesman*, 13 June 1992.
25. Bijaya Kumar Panigrahy, *Geochemistry of Rock Weathering in the Upper Part of Mahanadi River Basin*, Ph.D. Thesis, IIT, Kanpur, 1991.
26. J.C. Brown, and A.K. Dey, *Mineral Wealth of India*.
27. Ajaya Mitra Shastri, *India as seen in the Bṛhat-saṁhitā of Varāhamihira*, Motilal Banarsidass, Delhi, 1969, p. 325.
28. D.V. Singh, and A.K. Biswas, A Review on Diamond and its Beneficiation, *Journal of Mines, Metals and Fuels*, November and December, 1970, pp. 1-11.
29. James Prinsep, and Rājā Kāli Kishen, Oriental Accounts of the Precious Minerals, *Journal of the Asiatic Society of Bengal*, August 1832, pp. 353-63.
30. Sourindro Mohun Tagore, *Manimala — A Treatise on Gems*, Calcutta, 1879. Two Parts — Part II, The Eighty-four Sungs (gemstones) pp. 901-33.

3

Non-Gem Minerals in Pre-Modern India

As a sequel to our comprehensive report on minerals and metals in Ancient India,[1] we submit a tentative account of the non-gem minerals in India during the Muslim Period, better styled as 'pre-modern' India. The production and usage of gem minerals or precious stones in pre-modern India has been discussed elsewhere.[2]

We would chiefly record the nineteenth-century accounts of British travellers and geologists who looked backwards and performed the role of historians. Before we proceed with itemised discussions on salt, saltpetre, reh, alteration products such as copper, iron and aluminium sulphates, sehta, etc., and the metallic ore deposits, let us emphasise on the severe paucity of the chronological data on the subject.

The early Muslim world could collect information on Hindu medicine, astronomy and mathematics, but very little about Indian mineral resources. Abu Raihan Alberunī (AD 973-1050) was acquainted with the properties of gems and minerals; he even measured their specific gravities. But he failed to gain any insight about Indian alchemy based on the mineralogical method, because the Indians kept the information concealed.

Irfan Habib wrote:[3]

> The Delhi sultanate lacks any description of its economic resources of the kind that Abul Fazl supplies for the Mughal empire; only a very incomplete sketch can therefore be offered of its mineral and craft production.

Even though the Salt Range mines in Punjab were mentioned by Yuan Chwang, there was no further mention about these during the Muslim period before Abul

Fazl's writings in 1595. The Sambhar lake salt was however mentioned by Minhaj Siraj[3] in 1259.

The paucity of the data during the Muslim period was partly due to the widespread secrecy adopted by the vanquished Hindu people who refused to accept Muslim domination as a settled fact. As late as 1595, Abul Fazl admitted the ignorance of the Mughal rulers regarding the zinc ore processing technology which had been operating for many centuries in Zāwar (near Udaipur), Rajasthan. Even the European travellers, meticulous in their writings, often found the veil of secrecy difficult to pierce.

The Muslim rulers could gain access to the local mineral resources and indigenous technology only gradually, and with considerable difficulty. Ṭhakkura Pherū, born sometime during the end of the thirteenth century, was a Jaina scholar as well as an assay-master in Alāuddīn's court. He wrote in his *Dhātu-utpattiḥ* (8) how the copper ore from Babera (Babai, 8 miles south of Khetrī) could be reduced to the metallic state. Pherū also wrote about ores of lead, tin, mercury, cinnabar, realgar, orpiment, cementation brass, etc.,[4] but not about reduction-distillation of zinc which was going on at Zāwar and mentioned in the contemporary account of *Rasa-Ratna-Samuccaya*.

Pherū mentioned gold being obtained from river-bed washings[4] and not any sub-surface mine. Gwalior iron ore was one of the surface mine products which was exploited to produce iron and steel, right from the days of king Bhoja of Dhār, Mālwā, to the Sultanate and Moghul periods. Marco Polo and other travellers described pearl-fishing and diamond mining in the south.

References to mineral production in India are only slightly more during the Moghul period. Abul Fazl mentioned about gem minerals but not their sources; casual references are there in his *Āin-ī-Akbarī* about metalliferrous ores producing metals such as iron, copper, lead, tin, bronze, brass, etc., also salt works of Punjab and saltpetre.[5] The last one in the words of Abul Fazl, 'produces the explosive heat in gunpowder and used by his Majesty (Akbar) as a means for cooling water'. It was the former kind of use of saltpetre which attracted the attention of the Western colonial powers who started importing the material from India ever since the beginning of the seventeenth century.

The extensive trade of Indian saltpetre for several centuries, reported by

Moreland,[6] is hardly mentioned in any Indian writing. Similarly, the diamond production in India and its trade are subjects to be gleaned only from the accounts of the European travellers. Finally, the eighteenth and nineteenth-century visitors and geologists from England left accurate written accounts on minerals in India which have been beautifully summarised by Valentine Ball.[7] We would quote their reports on the ancient workings only, leaving out references to 'fresh discoveries' of mineral deposits in the nineteenth century, which are not relevant for our purpose.

Salt in India

Valentine Ball wrote[7] that 'there are several very distinct sources of salt in India, each of which is practically inexhaustible'. He referred to many coastal areas where salt must have been obtained since time immemorial by the evaporation of sea-water. Specific sites in Madras, Orissa, Bengal and Bombay coasts were mentioned. In the Chingleput district for example, manufacture by solar evaporation in pans started in January as soon as the rains were over. The process of evaporation was conducted for about a month. Complete evaporation was avoided to leave out magnesium sulphate and the hygroscopic salts in solution. Backwater and saline earth were widely exploited in India for recovery of salt, but these practices were often legislated against by successive rulers.

Alberūnī had made a remarkable postulate that a large portion of India had one time (in the early geological era) been under the sea. The British geologists confirmed this postulate by citing examples of saline lakes in Rajasthan and the salt ranges in the north-western part of the sub-continent. Eliot James wrote in 1880:

> The whole country of India, from Jhelum to the Afghan frontier is full of salt, and is constantly mentioned by English writers from comparatively early times.[8]

The Sambhar lake is situated on the eastern side of the Arāvallī range in Rajasthan, on the borders of the two erstwhile princely states: Jaipur and Jodhpur. Its greatest length is 20 miles, and the average breadth 5 miles. Its greatest depth being 3 ft, this lake is merely a hollow in the middle of long ridges of sand. The high salinity (never less than that of the sea-water even during the rains, sp. gr 1.03 at that time) after February (sp. gr. exceeding 1.20) results in deposition of salt. In hotter weather, the deposit presents 'a dazzling roseate-white efflorescence with blue, white and red crystals'. Left in the sun, there is some hardening of the deposit and the reddish hue is lost.

We have earlier referred to Minhaj Siraj writing on Sambhar lake salt in 1259. Production and distribution of salt were taxed and levied throughout the Muslim period. During the British period also, this mineral was often the bone of contention between the ruler and the ruled. Between 1856 and 1873 the Sambhar lake produced on the average 25,000 tons of salt per year which were distributed to the outside world by bullock-carts and camel-carts. In 1879, the production shot up above one lakh tons partly on account of the British management and partly due to the heroic efforts of the Rājpūtānā State Railway which was asked to, but could not handle an even larger tonnage! Ram Singh, the Mahārājā of Jaipur, requested the then Viceroy not to take away from him a part of his salt-revenue: "Please do not mention the word salt!"[8]

Salt reserves in Punjab

The province of Punjab in pre-partition India possessed enormous deposits of rock salt; the region still does. In the trans-Indus region there are deposits in Kohāṭ, and there are also deposits close to Kālābāgh, the latter being merely a continuation of those in the Cis-Indus Salt-range, geologically much more ancient than the Kohāṭ salt. A report on these deposits and their ancient workings has been provided by A.B. Wynne and Dr. Warth (*Memoirs Geological Survey of India*, Vol. XI, p. 299; Vol. XIV pp. 274-84) and quoted by V. Ball.[9]

The Kohāṭ district area of 60,000 square miles was estimated to possess over 40 million *mauṇḍs* of salt. In this region, the salt used to be obtained by open quarrying, and not by mining as in the salt range. The principal quarries were at Malgin, 'worked from time immemorial', at Bhadur Khel, 'opened twelve centuries ago' and at Jatta opened in the year AD 1650. Yuan Chwang's reference to Punjab salt mining could have been with reference to Malgin, Bhadur Khel or the Cis-Indus Salt Range itself. Jatta was developed during the closing years of Shāh Jahān. The Sikh control of some of the mines started during the latter half of the eighteenth century.

At Malgin, the salt crops out for a distance of three miles. The quarries at Bhadur Khel were four miles in length and about half a mile in breadth. At Bhadur Khel as well as at Kurruk, salt used to be obtained in 13"×13"×4" (thick) slabs (called tubbs) detached from the rock by pickaxe and wedge, utilising the natural stratification in the rock.

The Salt Range mines such as those near Piṇḍ-Dādan-Khān, about 100 miles north-west of Lahore were exploited for centuries. Akbar was informed about this range by one Asp Khan.[10] One Mīr Abul Qāsim was assigned by Akbar to administer the 20 *kos* long *namaksār* or the salt-range; he was known as *namakīn*. The merchants paid to the miners maximum 1/20 th of a rupee per *man* of salt and about the same amount of money for government levy, and sold the same in the Lahore market at about 3/8th of a rupee per *man*.

Ancient Sikh period workings were found at Khewra (subsequently expanded as Mayo mines), at Varchā, at Jutana, Sardi, Makrach and several other areas. Lieutenant Burnes visited the mine at Keoru, five miles from Piṇḍi-Dādan-Khān and left an interesting account (*Journal Asiatic Society of Bengal*, April, 1832, pp.145-8):

> We entered a cavern of irregular dimensions. The cavity has been excavated from the roof downwards. The salt is hard and brittle, so that it splinters when struck with the sledge-hammer and pickaxe. The rock is never blasted with gun-powder, from fear of the roof falling in; and accidents of this kind sometimes happen in the present simple mode of excavation. The mines are not worked for two months during the rains for the same reason. The miners live in villages among the hills. They receive a rupee for every 20 maunds of salt brought to the surface. The salt is not pure, having a considerable mixture of some substance (probably magnesia).
>
> 2500 maunds (1 maund equalling 100 English pounds) of salt are extracted daily, which gives about 8 lakh of maunds for the year. The price has gone up from a quarter of a rupee to two rupees per maund, exclusive of duties.
>
> It is closely monopolised by the Punjab (Sikh) Government, and Ranjit

> Singh hopes to derive an annual revenue of 16 lakh of rupees with 2½ additional for the duties. 1½ lakh of rupees is expended in working the mineral. The profits amount to about 1100 per cent, though the salt is sold for one-third the price of that of Bengal, which averages 5 rupees per maund of 80 lbs.

Eleven years later (1843), one Dr. Jameson visited the old Jutana mines, which had been operating for 20-35 years. The descent into the body of the mine was accomplished by steps cut in the salt, and the workings seem to have been large, but as irregular as usual in the Sikh excavations. The old workers at Varchā and elsewhere left the roof unsupported, which caused lot of roof-collapse. When the British experts like Dr. Warth took over, appropriate supports were provided in the modernised mines (such as Khewra renamed as Mayo) to take care of 'the dangerous chambers left by the old Sikh workmen'.

White modernising salt industry in India, the British rulers followed the Muslim, Marāṭhā and the Sikh rulers in controlling the salt trade and revenues, their actions being often prejudicial to the interests of the common men.

Alkaline salt, *reh* in India

Reh is the Indian term applied to efflorescent salts which have accumulated in the soil and subsoil waters of large tracts in India, often to such an extent that cultivation became impossible. The alkaline constituents of sodium carbonate and bicarbonate (apart from variable proportions of chloride and sulphate) however aided ancient and pre-modern civilisations in making glazed potteries such as NBPW, glass, soap, recipes of bleaching, washing, dyeing, curing hides for tannery, medical formulations, etc.

The efflorescent salt deposits have been variously known as *reh* in Uttar Pradesh, *khārī* in Bihar, *kallar* in Punjab, *ḍalla* (red in colour) and *papaḍī* (white) in the Vidarbha area, *chavuḍu* in Andhra Pradesh, etc. *Saji maṭṭī* is the name of the marketted crude mixtures of sodium salts lixiviated or leached out of *reh* and concentrated. Typical compositions of *reh*, *saji maṭṭī* and salt lake bittern are provided in Table 3.1.

Table 3.1
Typical compositions of *reh, saji maṭṭī* and bittern

(Taken from La Touche *et al.*,[11] Auden *et al.*,[12] and Brown & Dey [13])

Location	*Moradabad*	*Meerut*	*Kanpur*	*Sambhar Lake*	*Lonar Lake*
Material	*Reh*	*Reh*	*Saji Maṭṭī*	*Bittern*	*Incrustation*
Composition P.C.					
Na_2CO_3	9.66	4.68	27.02	14.5	8.2-33.2
$NaHCO_3$	8.55	3.96	4.28	9.1	6.3-24.4
NaCl	1.92	3.88	—	28.6	28.0 (av.)
Na_2SO_4	8.66	1.19	33.98	40.5	2.0 (av.)
K_2O	—	—	—	—	4.3-10.3

Auden *et al.* estimated[12] that the *reh*-infested districts of Uttar Pradesh alone produced 0.5 m ton Na_2CO_3, 0.6 m ton $NaHCO_3$ and 0.3 m ton Na_2SO_4. Similar data from other states have been compiled by Brown and Dey.[13] These are twentieth-century data which do not concern us. However, these illustrate the possible extensive use of the naturally occurring materials in the pre-modern India for which some information were collected and may be presented below.

Dr. Heyne mentioned in his *Tracts* (p. 45) that among Cītaldurg hills in Karnataka, an abundant commodity called *sabboo*, a mixture of sodium carbonate and sodium chloride, was sold in the local market. Captain Campbell noted similar material in the Salem district being recovered, from the soil resting upon *kaṅkar*, by a simple process of lixiviation and evaporation. The Lonar lake in Berār (Vidarbha) was noted by Capt. Mackenzie (*Berar Gazetteer*, p. 24) to be a crater-like hollow, more than a mile in diameter. The deposited salts (the range of analytical values provided in Table 3.1) used to be raked up by divers and sold as *ḍalla* (red, more of carbonate and bicarbonate) at Rs. 75-125 per kandy (500 lbs.?) and *papaḍī* or *paprī* (white, more of chloride) at Rs. 18-25 only.[14]

In 1832, R. Everest reported the deposits of salt efflorescing on the surface in the Bharatpur district and Bundelkhaṇḍ country which were being scraped off and used by the local inhabitants (*Journal of Asiatic Society of Bengal*, April 1832, pp. 149-50). The material consisted of sodium chloride, carbonate and sulphate. In 1834, J.S. Burt described 'The Mode of Extracting Salt from the Damp Sand-beds of the river Jamunā as practised by the Inhabitants of Bundelkhaṇḍ' (*Journal of the Asiatic Society of Bengal*, January 1834, pp. 33-6). Burt found a process of lixiviation or leaching of the *phultā* (swelling-up) sand in operation on the Jamunā river bed 24 miles from Kālpī. After lixiviation, the salt extract was boiled in earthen pots until it evaporated leaving behind a salt at the bottom. From 80 lbs of sand, about one lb of salt was obtained. The unrefined salt was sold at one rupee per maund of 40 seers or 80 lbs. Burt also noticed people living on the Ganges at Kanpur scraping the ground up to half an inch and extracting the salt by lixiviation.

J. Stephenson reported (in the same issue of the Journal, *JASB*, January 1834, pp. 36-7) from Ghāzipur that the salt from that locality contained not only sodium chloride, sulphate and carbonate but also nitrate of potash and lime. In the village of Ratouly there was a thick layer of *kaṅkar* impregnated with a large proportion of saline matter, chiefly chloride, sulphate and carbonate of sodium. The formation of such a proportion of saline matter was presumably on account of evaporation and interaction (with limestone in *kaṅkar*) of the sub-surface and seepage water. The famous French chemist Berthollet provided similar postulates for the formation of Natron lakes around the river-beds of Egypt.

Borax or the hydrous sodium bi-borate also occurs as an exudation on the surface of saline soils and often in connection with salt lakes and associated with sodium chloride and sulphate. It has been widely known and used in the ancient world: as *tankana* in Sanskrit, *tshale* in Tibetan and *tinkal* in the Western world. India's chief supply has been from the hot springs in the Puga (or Pugha) valley of Kashmir (known as *sohāgā*) and Gnari or Hundes and the series of lakes such as Chapakni, Rudokh, Tengri-Nur, etc., in Tibet. Ball quoted E.T. Atkinson describing the Tibetan process[15] for extraction of borax. After lixiviation and evaporation, the crude borax was further refined by pounding, removal of impurities in lime-water, filtration, dissolution in boiling water and then subsequent evaporation and crystallisation in *kānsā* (alloy of copper, zinc and lead) vessels. There are Dutch records of 1618-19, 1625, etc., showing export of Indian borax from Masūlipaṭanam.

Saltpetre in India

For many centuries the entire world was dependent for its supply of saltpetre, essentially potassium nitrate, also containing some sodium and calcium nitrate — on superficial deposits resembling *reh* which has been described before. Whereas formation of non-nitrogenous *reh* has an entirely geological basis, biological factors have played crucial roles in the formation of *sorā* or saltpetre, which is nitrogenous. The bio-geological mechanism of formation of saltpetre deposits in India was discussed for the first time by W.J. Palmer, whose paper (*Journal of Chemical Society*, London, Vol. XXI, 1868, p. 318) was extensively quoted by Mallet.[16] Nitrogenous refuse like urea, ammonia, etc., are converted to nitrite and then to nitrite by nitrosifying and nitrifying bacteria respectively.

The saltpetre-yielding Indian soil usually had deep beds of *kaṅkar* or granules of carbonate of lime, a source of potash from felspar or wood-ash thrown on surface drains, bearing discharge of nitrogenous refuse such as food-waste and human or animal urine. The ancient *sorā-wāllāh* or the saltpetre-hunter would look for 'the white veil-like patch of crystalline formation' near the surface drains of human or animal habitations. He would scoop these surface earth samples and lixiviate in boiling water in earthen vessels. Subsequently, the filtered solution would be evaporated up to a point, and then cooled to yield the first crop of crystals chiefly of potassium nitrate. The mother liquor on further evaporation yielded calcium and sodium nitrate and chloride. Whereas potassium nitrate is not deliquescent, and this dry material separated from the rest has been widely used for many centuries as constituent of gunpowder, the entire material including the hygroscopic residue, when dissolved, cools water and was widely used for preparing cold drinks.

As Palmer emphasised, the formation of saltpetre depended upon natural sources of lime, carbonate, potassium/sodium and organic nitrogen simultaneously, and deposits were available all over the world. Sarton has outlined the history of saltpetre and gunpowder in the medieval world.[17]

There was a saltpetre industry in al-Basra in the ninth century AD, and

Muslims knew potassium nitrate as *bārūd*, but it is doubtful whether 'gunpowder' was invented earlier than thirteenth-fourteenth century. Roger Bacon provided a secret recipe in his Tract of 1242, and al-Hasan al-Rammah described (around AD 1280) the purification of potassium nitrate by fractional crystallisation, separating it from the hygroscopic impurities left in the mother liquor. This purification was the crucial step in making good quality saltpetre which gave gunpowder in admixture with charcoal and sulphur. The first actual use of saltpetre as a constituent of gunpowder was made in the lower Rhine around AD 1320-30.

P.K. Gode's review (*Studies in Indian Cultural History*, Pune, 1960, Vol. II, pp. 1-56) suggests that gunpowder was probably not used in India before the fifteenth century.[18] The earlier Sanskrit references to the fire missiles (*Yogavārttikā*) probably corresponded to arrows or darts tipped with inflammable substances.

During AD 1132-64, China invented bamboo tubes filled with pyrotechnic composition of saltpetre, sulphur and charcoal to throw catapults from a distance. Cast iron cannons were available by AD 1356 (*Journal of Chemical Education*, Vol. 24, November 1947, pp. 522-37). This technology diffused to India. Mahaun, a Chinese, visiting Bengal in 1406, reported availability (there) of 'knives, scissors, steel and guns' (Philips, *JRAS*, 1895). Towards the end of the century, Gajapati Pratāparudradeva became ruler of Orissa (AD 1497) and wrote *Kautukacintāmaṇi* which contained references to sulphur, charcoal, saltpetre, *veṇunāla vaṁśanāla* or hollow piece of bamboo through which catapults could be thrown.

The reported use of pyrotechny by the Vijayanagara Kings (~ AD 1443) and the Persian treatise on fireworks (AD 1466) by Zain-ul-Abidīn, a ruler of Kashmir suggest that the technology might have been also introduced by visitors and invaders from the West. On 1st November 1526, Bābur witnessed the casting of a gun. In 1528, he used cannon and gunpowder near Kanauj.

Abul Fazl in his writing of late sixteenth century gives a crisp account of the twin use of saltpetre:

> Saltpetre, which in gunpowder produces the explosive heat, is used by his Majesty (Akbar) as a means for cooling water, and is thus a source of joy

> for great and small.
>
> Saltpetre is a saline earth. They fill with it a perforated vessel, and pour some water over it, and collecting what drops through, they boil it, clean it, and let it crystallize.
>
> One *ser* of water is then put into a goglet of pewter, or silver, or any other such metal, and the mouth closed. Then two and a half *sers* of saltpetre are thrown into a vessel, together with five *sers* of water, and in this mixture the goglet is stirred for about a quarter of an hour, till the water in the goglet becomes cold. The price of saltpetre varies from ¾ to 4 *maunds* per rupee.[19]

Abul Fazl also mentioned the profitable trade of bringing ice from the mountains to the Lahore market by the river route. We would see later that Indians probably manufactured ice using the cooling property of saltpetre when dissolved in water.

The Portugese were already firmly established in India when Bābur started on his great adventure. Both the Western colonialists and Moghul invaders were conversant with the use of gunpowder. The seventeenth-century saltpetre export trade (from India) 'must be sought in the military history of Europe', in the words of Moreland. The defeat of the Spanish Armada in 1588 brought in the ascendancy of the Dutch and then the British power and their constant warfare in several continents.

> By the beginning of this period (early seventeenth century), the progress of war was outrunning the European supplies of saltpetre, and the discovery that practically unlimited quantities of saltpetre were available in India was a very great advantage to those belligerent nations which were in a position to obtain it by the ocean route.[20]

The first reference that Moreland found to the export of saltpetre from India occurred in a letter from the King of Spain written in the year 1605, laying stress on the scarcity of saltpetre in Europe and directing the Viceroy of Goa to send home ten or twelve casks of saltpetre yearly, until further orders. The commercial initiative for manufacture of saltpetre using Indian material and cheap labour was taken by the Dutch in the Coromandel coast sometime during 1617-21. The English company followed the Dutch example.

It seems that the Moghuls did not favour the export of saltpetre which had strategic importance, and the Western nation companies often arranged despatch of saltpetre as 'sugar'. Even this was hazardous during the famine year of 1630 when the foreign companies issued a warning to their employees 'not to pass saltpetre through the inland customs as sugar, lest it should be detained as provisions and the fraud be discovered'.[21] In the year 1646, 'Prince Auraṅgzeb, then Viceroy of Gujarat, forbade export of saltpetre on the ground that gunpowder made from Indian saltpetre might be used against some Muslim power, but the transfer of the prince to another government quickly removed the obstacle'.[22] Whereas Jehāngīr and Shāh Jahān were more interested in Western toys, Auraṅgzeb paid more attention to the threat from his brothers rather than the threat from the emerging Western powers and their trade in Indian saltpetre which continued to be exported. However, a new toll was imposed and the Moghul emperor made saltpetre an 'imperial monopoly' in the year 1655.

Saltpetre was obtained from various parts of India. The Coromandel coast was the first region to be exploited; a little later Gujarat, Agra and some of the Konkaṇ ports came into the list. However, the Bihar product exported through the Orissa ports and then Hooghly, gained paramount importance. The cost of a *mauṇḍ* (of saltpetre) of 74 lb at Patna was about the same as for a maund of 37 lb at Ahmedābād, or just half the price.

Between 1631-50, the annual Dutch exports ranged between 200 and 300 tons; around 1661 it shot up to 1480 tons, largely on account of the Dutch investment and improvement of the saltpetre technology. The total output in the Patna area around 1688 came to 226, 200 *mauṇḍ*s raw and 127, 238 *mauṇḍ*s refined.[23]

The larger output as well as better quality of the Patna saltpetre was due to the Dutch involvement in the technology. It is doubtful whether Indians themselves had realised the need for refining of Indian saltpetre. Calcium and sodium nitrate, unlike potassium nitrate, are hygroscopic and ruin the property of gunpowder. On storage, the unrefined material absorbs moisture causing particles to agglomerate, further jeopardising the kinetics of its reaction with sulphur, etc., and its firing property. In the 1620s, the English factors in India complained about unrefined Indian saltpetre which 'will not keep, for being ill-corned it grows all into clods', and that it was necessary to refine the 'heavy ancient-looking powder'. Such practical experience must have resulted in coining the famous proverb that 'one should keep his gunpowder dry'. The solution to this problem obviously lay in fractional crystallisation of potassium nitrate by cooling hot saturated nitre brine, leaving the more soluble and hygroscopic salts in the liquor.

The Dutch must have tried refining Indian saltpetre as early as 1621, since an invoice of theirs from Masūlipaṭaṇam dated October, 1621 notes despatch of 47,500 lbs of saltpetre and 8,700 lbs of 'refined' saltpetre.[24]

The Indian process of the manufacture of saltpetre involved boiling of lixiviated and filtered solution of the nitre earth in earthenware pots. The English factors at Madras complained in 1652 that this is a 'tedious and troublesome' process and for having 600 bags of saltpetre they had to witness 200 earthen pots 'falling into pieces'.[25]

The Dutch had already imported copper pans from Holland and set-up a refinery in 1640-1 at Pīplī in Orissa.[23] When they imported similar metallic boilers for their works at Chaprā, near Patna, the local Indian *nuniā*, who were competitively making saltpetre, using earthen pot for boiling the nitre brine, felt worried at this competition, and stopped supplying whey (to the Dutch) which was an essential material for bleaching saltpetre.[26] Tavernier witnessed and reported this episode. It is significant that the local rulers were interested in levying tax but not improving the technology, and the poor villagers had no capital for fabrication or import of copper vessels.

Tavernier observed that the gunpowder made in Bhutan was of long grain and that made in Assam was 'not as long but round and small like ours, and is much more effective'. What procedures were adopted at Bhutan and Assam for purifying saltpetre is not known. The powder had to be dry as well as fine. The typical Indian powder-houses were mat or straw sheds where women and boys did the grinding by hand and foot with wooden pestles and mortars, and the ingredients were shifted in leather sieves. Even the English factories in India were following 'the crudest method of native powder-makers'. Only after the 1740s, machineries came to India from England: rolling and pilon mills, iron cylinders for grinding saltpetre, copper and iron boilers. In 1754, mechanical grinding mills were drawn by buffaloes and 1863 saw the first steam mill in saltpetre technology.[25]

Ever since the middle of the eighteenth century, the East India Company of England enjoyed the monopoly of Indian saltpetre[26], a very powerful ammunition targetted against the land which produced it! During the middle of the nineteenth century, the British government allowed some European firms and even some Indians to handle the saltpetre trade.

Valentine Ball reported[27] in 1881 that during 1876-80 more than 1.67 million hundred weights (cwt) of saltpetre valued at nearly 1.37 million pounds (£) were exported from Calcutta. The average price was around 16½ shillings per cwt of 112 lbs. Two-thirds of the export through Calcutta came from the districts of Tirhut, Sāran and Champāran, the rest from Kānpur, Ghāzipur, Allāhābād and Vārāṇasī in the Uttar Pradesh. The districts of Madurā and Nellore in south India were also producing saltpetre. The saltpetre manufacture in India declined on account of the rising cost of production and cheaper sources outside the country.

Cooling water and other drinks was one of the major uses of saltpetre in medieval India. The heat of dissolutions (Δ H) in water of most of the constituents of saltpetre are endothermic as evident from the thermodynamic data in kilo calorie/gm mol of the solute, all the values having negative sign (absorption of heat, resulting in cooling):

KNO_3 8.633, $NaNO_3$ 5.05, $Ca(NO_3)_2$, $4H_2O$ 7.99, $CaCl_2$, $6H_2O$ 4.11
KCl 4.4, K_2SO_4 6.32, Na_2SO_4, 10 H_2O 18.74, etc.

It is noteworthy that anhydrous $CaCl_2$, $Ca(NO_3)_2$, Na_2SO_4 dissolve in water evolving heat, i.e., dissolution is exothermic; but the dissolution of hydrated salts is endothermic. Thus hygroscopicity of some of the constituents of saltpetre, hindering the efficacy of gunpowder, actually aid in its cooling property.

We have earlier mentioned Abul Fazl's recipe for the coolant bath water: 2½ seer saltpetre in 5 seer water (equivalent to 50 g saltpetre in 100 g water) which is about the solubility limit of potassium nitrate in water at 35°C. Using the thermodynamic data (Δ H), it can be shown that the outside bath temperature should drop by 20-25°C which should cool the water inside a separate container lowering its temperature by 10° C or more. This process could have been repeated through successive stages of cooling. At 0° C, solubility of KNO_3 in water is 13.3 g/100 g and it can be shown by calculation that one kg of saltpetre should abstract heat equivalent to freezing of one kg of water to ice.

We are not sure whether the above-mentioned possibility was actually executed in pre-modern India. Dharampal has quoted[28] an AD 1775 paper written by Sir Robert Barker who witnessed 'the process of making ice' at Allāhābād (December night) where the atmospheric temperature did not seem to be sub-zero in degrees centigrade. Boiled (and then cooled) water was exposed to the atmosphere in shallow trays. The use of saltpetre was mentioned in the paper but in a different context, viz., it was mixed with ice to provide sub-zero temperature in the external bath so that ice-cream could be congealed in an inner silver cup. It is quite possible that very cold water collected at the end of December night was mixed with saltpetre to provide sub-zero temperature bath and used to prepare ice within a container. The water to be frozen was previously boiled and cooled. This prior treatment helped, probably on account of precipitation of the constituents in the hard water as solutes adversely lowered the freezing point.

Mining and smelting of metallic ores in pre-modern India

We have reported mining and metallic ores in ancient India elsewhere.[29] Circumstantial evidences suggest that Rajasthan provided the earliest copper ores to chalcolithic India. Rajpūra-Daribā near Udaipur provides the earliest C-14 dated mine: 1260 ± 160 BC, the ore-body having Cu-Zn as well as Pb-Ag ore lenses. Tabulation of the various C-14 data[29] show exploitation of Rājpura-Daribā mine between 1260-1000 BC, and then again 375 BC-AD 110, Rampurā-Aguchā Cu-Zn and silver-lead mine 370-200 BC, Zāwarmālā zinc-lead-silver ore mines 430 BC to AD 60 and continuously in the post-Christian era, Hatti gold field, Karnataka 760 BC, etc.

Some of the medieval era C-14 dated mines are listed in Table 3.2. Some of the undated copper ore mines, listed in Table 3.3 must have been exploited in the pre-modern era.

Copper ore mining

Archaeological evidence suggests that there were a number of prosperous towns flourishing in Rajasthan during the Historical period (400 BC-AD 400) close to the copper mining areas (also yielding zinc, lead and silver); many of these are unfortunately undated.

Ambamata, Gujarat mine (140 BC onwards) was exploited during the Śuṅga period for copper, lead and zinc ores. Ingaldahal, Karnataka copper mines yielded the ore for the periods AD 170-220.

The *Seraks* or lay Jains initiated copper mining in the Singhbhūm and Hazarībāgh districts during the Historical Period, and the tradition persisted till the sixth century AD. Late Kuṣāṇa type copper coins have been found in the Rākhā Mine. The Camolī and Almorā district mines in Uttar Pradesh and the Kulu (Himachal Pradesh) mines were also active during the sixth century AD and later periods. In the ninth century AD, the copper mines of Nepal became famous and the product received eminence as *Naipālī* copper.

Table 3.2
Some post Christian era C-14 dates of Indian mines of antiquity

Date	*Site*	*Comments*	*Reference*
AD 150	Rājpura-Daribā	Cu-Pb-Zn East Lode 154 m depth	Willies[30]
AD 110 ± 130	-Do-	PRL 210	Chakrabarty[31]
AD 60 ± 50	Zāwarmālā	Zn-Pb ore Timber from Launder BM 2149	Paliwal *et al.*[32]
AD 30 ± 70	Hatti, Karnataka	Gold Ore. Site No. 1	Agrawal *et al.*[33]
AD 110 ± 70	-Do-	Gold Ore. Site No. 2	-Do-
AD 475 ±115	Kolar, Karnataka	Gold Ore. Champion Reef. 50 m depth. Sample TF 879	-Do-
AD 685 ± 90	-Do-	Gold ore. Superficial Excavation Sample TF 1199	-Do-
AD 170 ± 35	Ingaldahal Karnataka	Copper mine. 30-40 m depth. Wood Sample BM 2364	Paliwal *et al.*[32]

AD 220 ± 110	-Do-	PRL 252	Chakrabarty[31]
AD 1130 ±100	Ambamata, Banaskantha, Gujarat	Copper-Lead-Zinc Ore Bore-hole A 100, PRL-66	Agarwal *et al.*[34]
AD 1070 ± 85	Kumbaria, Banaskantha, Gujarat	Copper-Lead-Zinc Ore TF 1222 charcoal	-Do-
AD 1440 ± 90	-Do-	TF 1221	-Do-
AD 1070 ± 80	Bandlomattu Hill, Guntur, A.P.	Copper Ore TF 805	Agarwal *et al.*[34]
AD 1320	-Do-	TF 806	-Do-
AD 1440 ± 90	Mailaram, Khammam, A.P.	TF 373	-Do-
AD 1640 ± 80	Kaladgi, Hassan, Karnataka	PRL 254	-Do-
AD 1750 ± 60	Zāwarmālā, Rajasthan	Zinc smelting site charcoal BM 2223	Paliwal *et al.*[32]
AD 1810 ± 50	-Do-	Balaria mine timber BM 2338	-Do-

Table 3.3
Some undated copper mines of aniquity

Name and Location	*Comments*
Rajasthan	
1. Ganeshwar area mines, e.g., Chiplata, Baleshwar, Daribā etc., near Neem-Kā-Thānā	Pre- and Mature Harappan workings and connected with Jodhpura culture (2800-2500 BC).
2. Baldeogarh, Alwar	10 m deep, 300 m length, large heap of copper slag.
3. Daribā, Alwar	75 m deep, 400 m length. Chalcopyrite with Ni and Co. Nearby river Sota named after cobalt sulphide mineral.
4. Bagor, Jhunjhunu	Old workings. Central shaft of once prosperous mine.
5. Ajītsāgar, Jhunjhunu	Reported to be one of the largest mining centres in the past.

6. Dhanaotā, Jhunjhunu	35 m depth. In operation till the end of the nineteenth century.
7. Khetri Copper belt 80 km from Singhana to Raghunāthgarh	Mining since the Maurya period. Many operations during AD 1590-1895. 0.7 m ton slag in Singhana.
8. Daribā, Udaipur	Water-logged. Load multi-metallic. One portion shows earlier C-14 dating 1260 ± 160 BC PRL 208 quoted by Chakrabarti.[31]
9. Chanpurā near Rakhabdeo, Udaipur	Extensive old working for copper.
10. Delwārā-Keroli area	Ancient copper mining at Vilota and smelting at Pipliā, Delwārā.
Uttar Pradesh	
11. Pokhrī, Camolī district	One of the most reputed localities for ancient copper mining. Of the four old workings, the Raja's mine used to produce 300 kg ore assaying 25 per cent Cu per day.
12. Thela, Almorā	A number of old copper workings. Quartz, malachite, dolomite rejects in dumps.
Bihar	
13. Rākhā Mines, Singhbhūm	50 m depth, 122 m length. First six post-Christian centuries. Kuṣāṇa coins
14. Mosabani, Singhbhūm	Old working among the largest met within the belt. 20 m depth. Some rotten timbers found in 1934 not carbon dated.
15. Baragaṇḍā (12 × 4 = 48 mines) Hazārībāgh	Extensive old workings contemporaneous with Singhbhūm copper belt. 1200 m length 4 m width. Slag heaps.
Andhra Pradesh	
16. Agniguṇḍalā, Guṇṭūr, A.P.	Open cast mining, long trenches 2 miles, reaching depth of 30 m. Copper, lead and also silver. Nearby Bandlomattu mine C-14 dated AD 1070-1320

The C-14 dates of AD 1070-1440 (Table 3.2) show the activities of the copper-lead-zinc producing mines of Bandlomattu — Agniguṇḍalā area in Andhra Pradesh, Ambājī and Kumbariā in Gujarat and Deri mine in Rajasthan.

At Agniguṇḍalā, the old workings are spread over a linear belt of more than two miles and there were rock dumps, ore dumps, pounding stones, washing tanks, tailing dumps, smelting furnaces, crucibles or cupolas and evidences for extraction of copper and lead.[35] The nearly Bandlomattu mine has been C-14 dated as AD 1070-1320.

In the Ambājī and Kumbariā (24° 19-20'N. 72° 51'E) areas of Gujarat and Deri (24° 23'N. 72° 50'E) area of Rajasthan, the mining and metallurgy of copper, and possibly also of lead, zinc, gold and silver, commenced in the second century BC. Evidence for the continuity of activities during Kṣatrapa period of second to fourth century AD and again intermittently from early eleventh century AD, corresponding to Solaṅkī period in the earliest part, are also available.[36]

The present village of Kumbariā was formerly known as Arāsān, derived from *Ar* or metal. In the book *Sri Arasan Tirth*, Vishal Vijayji (1961) has described how 'the treasure of metals near Arāsān' was utilised by the Solaṅkī dynasty and subsequent rulers (AD 1030-1619), and the wealth used in constructing a series of temples at Kumbariā, Delwārā, etc.[36]

Pherū's reference to copper smelting at Babai during the end of the thirteenth century has been already cited.[4] Abul Fazl wrote in his *Āin-ī-Akbarī* (AD 1590) that "Babai has a stone fort and a copper mine. Singhana has a copper mine and mint for copper coinage." Subsequent to this record, there appear to be no documents on the mining activity till the beginning of the nineteenth century. Monetary records analysed by Irfan Habib indicate that between 1595 to 1614 and even beyond, the price of copper (relative to silver) in India went on ascending indicating greater demands for use in bronze cannons, vessels and utensils.[37] The Ahom state in Assam used to import copper from Moghul India and export bell-metal vessels,[38] tin coming presumably from Myanmar and Malaysia.

Our chronological data for the manufacture and use of copper in medieval India are indeed scanty, and now we turn over to the nineteenth century records of British travellers related to copper mines in India. Valentine Ball presented a beautiful summary[39] of these records.

The ancient copper mines in the Nellore district situated between Nellore and Gunṭūr were first reported by Dr. Benjamin Heyne in his *Tracts, Historical and Statistical in India* (1800). Later C.A. Kerr described that 'the ancient excavations are of prodigious magnitude, many of them occupying several hundred feet square, and having a depth of 50 or 60 feet'. James Prinsep analysed some of the copper ores from Nellore (*Journal Asiatic Society of Bengal*, Vol. 2, Feb. 1833, p. 95 and Vol. 4, Oct. 1835, pp. 574-84).

Captain Newbold gave a detailed list of the ancient copper mines in this area (*Journal Royal Asiatic Society*, Vol. 7). He was told by the residents of the site Garimanipenta or Gunnipenta (lat. 14°59' 30"N, long. 79° 37'E) about the tradition that the local mines were first worked by the Vijayanagara kings. The principal ancient native mines were 30 miles further north, near Gorganpully, where the mines were 100 feet deep and several 100 feet long.

The ore variety was interesting: basically carbonates and silicates of copper. A 'new mineral *mysorin*' was later proved by F.R. Mallet to be an impure malachite, owing its dark colour to admixture with ferric oxide and chalcocite (Records *GSI* Vol. 12, p. 171).

Valentine Ball reported a large number of ancient copper mines in Bihar which unfortunately stand undated (*vide* Table 3.3 Nos. 13-15). A few must have been exploited as early as the Kuṣāṇa period. The mines around Badia and Surda were surrounded by huge slag heaps indicating widespread smelting. A cluster of 48 mines in Baragunda (lat. 24° 4' 3" N, long. 86° 7' E), Hazārībāgh and large mounds of slag in the vicinity deserve special mentioning.

Valentine Ball noted some details about the ancient mining of copper ores in Myanmar and Afghanistan but these are outside the scope of this article. Nepalese copper had been famous in India since eighth century, but Ball could not obtain any information on the subject on account of 'the jealous way in which the resources of that country are kept hidden from the public'. Medlicott visited Nepal in 1875, but was not permitted to examine the copper mines.

Mining and smelting in Sikkim and Bhutan were exclusively in the hands of the Nepalese. The ancient mines have been described as 'magnified rabbit holes, passages meandering with courses of the ore'. The tools used were iron hammer, gad or chisel. The ore pounded on the 'knockstones' outside the mine was washed by women in small tyes which looked like partitioned troughs. The washed ore was dried and then smelted in 18 inches deep furnace built of refractory clay. A typical furnace was one foot square at top, and tapering to the bottom, where charcoal was rammed down so as to form a floor through which the molten metal could sink (Percy's *Metallurgy*, pp. 387-90).

Captain Herbert's memorandum on the mines and mineral productions of the Kumaūn and Garhwāl districts of the Himālayas was published in 1829 (*Asiatic Researches*, Vol. 18, 1829, p. 239). In 1838, two mines in working condition — Rai mine in the Gangoli *parganā* and the Sira mine in patti Barabisi, in *parganā* Sira, both in the eastern part of Kumaūn — were examined by a Cornish miner who had been brought out from England by Colonel Drummond. The miner naturally found fault with 'the wasteful system, or want of system, practised by the miners', and suggested improvements in the tools; enlargement of the narrow burrows into practicable passages; the use of timbering, the introduction of pumping machinery, etc. (*Journal Asiatic Society of Bengal*, Vol. 7, November, 1838, pp. 934-41).

The matrix of the ore-body was stated to be a talcose slate with quartz, calc spar and dolomite. The cupriferous band was about two feet thick, of which 20 per cent was ore. Pulverising and washing of the ore was being done manually by women.

> In Cornwall, one woman provided with a wheel-barrow and shovel for the conveying and washing of the ores, and a boy with a sieve for dressing them, would accomplish a task equal to that of ten women on the (Kumaun mine) system. . . . The water raised by six men (drainage of the Sira mine) could be effected with a hand-pump by one man; in order to keep that pump constantly going, two men might be required, but the remaining four would be added to the number of those who are excavating.

The improvements suggested could not be excuted because the English copper was selling at Almorā at a price below that of Kumaūn copper which was unable to compete in the market, and production had to be stopped.

The Rajasthan mines had supplied copper ores ever since the Harappan era through the Historical Period and continued to remain active during the

Sultanate and Moghul periods. Some of the copper mines in the states of Alwar, Jaipur, Bharatpur, Ajmer, etc., have been mentioned in Tables 3.2 and 3.3. The most noteworthy amongst these were Singhana, Khetrī, and Babai in the Jaipur State, which belonged to the Rājā of Khetrī.

Mining and metallurgy in Khetrī Copper Belt

The Khetrī Copper Belt is located in the semi-arid tracts known as 'Shekawati country' in the north-eastern Rajasthan. The 80 km long belt extending from Singhana (28°06' N,75°54 E) in the NNE to Raghunāthgarh (27°30' N, 75°21' E) in the SW, is dotted with old workings, dumps of mine waste and slag heaps over many localities, bearing ample testimony to the ancient mining and smelting activity. The first report on the subject was written by Captain A.E. Boileau in April 1831 visiting the Copper Works at Singhana (*Gleanings in Science* No. 36, December 1831, pp. 380-4). Col. J.C. Brooke presented a similar report on the Kolihan Works at Khetrī (*Journal of the Asiatic Society*, Vol. 33, 1864, pp. 518-29).

The ore-body contained sulphides such as chalcopyrite, pyrite and also ferro-cobalite or *sehtā*, blue and yellow vitriols, alum, etc., which we would discuss later.

The openings of numerous closely spaced shafts were observed. When the shafts had reached a sufficient depth, galleries were driven out in various directions to follow the veins of ore. This gave a honey-combed structure and provided a good ventilated mine-system. Rocks were rent and split by fire-setting procedure.

> Each labourer is provided with a lamp, *diyā*; a hammer, *hātori*; a mining chisel or gad, *tankī*; and a small wicker basket, *ṭokrī*. The lamp is placed upon the workman's head, which enables him to watch the glittering particles, and detach such portions of the ore. While at work, he seats himself upon his heels, and the little basket upon his knees, in which he receives all the fragments of ore that are struck off by the chisel.

On account of the arduous work, the labourers had very short life betwen 35 to 40 years. They earned only two *ānnā*s (one-eighth of a rupee) a day for which they had to furnish one basket full of ore. The mining schedule was for eight months — between October and May prior to the onset of the rainy season. As the mines got deeper, dewatering of the mines had to be attended to. Due to tortuous course of the mine, the miners formed a human chain from the minehead to the gallery, passing pots (*ghurrahs*) filled with water. It was a slow and expensive process. In the Kolihan mine, 27 people had to be engaged for this purpose for which a monthly expenditure of Rs. 200 had to be incurred.

When the day's work was completed, the miners' produce were sold at a public *cabūtarā* near the smelting furnaces *mandāri*.

The proprietor of the furnace usually a Muslim Bhora, employed a person at rupees three per month to separate the ore from the schistos rock, and to break it into small pieces which were finally powdered by using 15 kg hammers. It was next mixed with cowdung and made into rolls about 10 cm long *piṇḍī* which were first dried in the sun and then roasted in a fire of cowdung cakes. The ore was then ready for the smelting furnace.

The potter who built and worked his own furnace had four assistants who often belonged to his family. The small (about 3½ ft high, 1 ft in diameter) blast furnaces were prepared in the following manner:

> A quantity of common sand was spread on the floor of a circular hut, in the centre of which a depression, 12 to 15 inches in diameter and 2 or 3 inches deep, was made; in this a layer of fine sand and another of ashes were laid to prevent the metal from adhering to the bottom of the receiver; two clay nozzles or tuyeres were then placed on opposite sides of this hollow, and a third between them, leaving the fourth side vacant for the slag to escape.
>
> The nozzles were then connected by moist clay, and a circular rim of mud, a few inches in height, was raised, on which three annular vessels of fire-clay were placed to form the body of the furnace; each of these (*kothi*) was 15 inches in external diameter, 10 inches high, and 3 inches thick. They were used repeatedly, but the lower part of the furnace had to be reconstructed for every charge.

> The bellows (*phonknī*) were simply goat-skins connected with the nozzles, and were worked by the families of the smelters. After a preliminary firing to dry the mud, the furnace was charged with charcoal, roasted ore, and iron slag or *rīt* the latter being employed as a flux. (Vide Plate 38)
>
> In a day of nine or ten hours' duration, 3 maunds of charcoal, 2½ of the roasted ore, and 2 of the iron slag were consumed. The slag was drawn off and the smelted copper which had accumulated at the bottom of the furnace was removed on the following day. It was then re-melted and refined in an open furnace (*guzāb*) under a strong blast from bellows, and cast into small bars or ingots which were subsequently removed to the Mint and cut up and fashioned into coins.
>
> The ore was said to yield only from 2½ to 7½ per cent of metal, but the profits must have been not inconsiderable, as the Rājā of Khetrī is said to have claimed one-sixth of the value of the copper in addition to Rs. 14,000 received for the lease. The lilac-coloured and brittle metal from Singhana is said to have been inferior to that of Basawar in the Bharatpur state, this being attributed to the use of the iron slag as a flux.[40]

The above summary on Singhana Works was prepared by Valentine Ball from the 1831 report of A.E. Boileau. The 1864 report of J.C. Brooke on the neighbouring Kolihan Works gave details which were naturally very similar. Brooke provided useful cost figures for smelting 5 maunds of ore: cost of ore, Rs. 5; 4 md Charcoal 1 Re. 4 *ānnās*; 5 md flux 4 *ānnās*; labour cost 11 *ānnās*; refining cost 8 *ānnās*; i.e., a total of 7 Rs. 11 *annas*. After paying royalty in the form of copper the remaining metal would fetch Rs. 9. Thus profit of the operation would have been slightly over a rupee.

The mining activities in the Khetrī area ceased around 1872 for several reasons. Mining conditions became difficult. There were lack of facilities for speedy dewatering of mines. The profitability of the operation, as indicated above, was poor, specially because 'the Jaipur Darbār had refused to reduce royalties'. The main reasons for the stoppage of the mining activity were that these mines could not stand in competition with the metal imported into this country and the prohibition imposed by the British Government in 1869 for the minting of copper coins. Thus the pre-modern technology which had been in operation for many centuries came to an end.

After attaining national independence, plans were afoot to start modern mining and smelting industry at Khetrī. The geological exploration work was started in 1954 and decade later, Hindustan Copper Limited started its project on the Khetrī Copper Complex. In connection with the recent exploration work, some more facts about the ancient mines have come to light.[41, 42] The ancient mines of Madhan-Kudhan, Kolihan, Akwali and Satkui were investigated.

At Madhan-Kudhan, 5 old workings over a strike length of two km were examined in detail.[41] The depth of these mines varied from 15-45 m. Four mines of the Kolihan Copper deposit over a strike length of 500 m. were examined.[42] The mine dump occurring near one of the mines was one of the largest (about 0.5 million tonnes). The largest heaps of slag found near Singhana village are estimated to be around 0.7 million tonnes averaging 0.3 to 0.5 per cent copper.

A rough estimate of the total quantity of copper ore mined in the past suggest a probable figure of 5 or 6 million tonnes from the mines located in the Khetrī area. Out of this quantity around 3.5 million tonnes appear to have been mined in the Madhan-Kudhan sections alone,[41] where the average copper content in the mine spoil is about 0.5 per cent. In Kolihan, a higher grade ore was mined. The maximum copper content analysed in the mine dumps is 0.8 per cent.

If we assume that 6 million tonnes of ore mined in this area during 300 years (1590-1895) assayed 2 per cent on the average, then more than one lakh tonne of metal must have been produced during this time. Raghunandan has paid a tribute to the ancient miners not only for their productivity but also for their skill in distinguishing between the leached and oxidised copper-bearing rocks and the non-copper bearing ores.[42]

Sehtā or ferro-cobaltite in Khetrī copper mines

The Khetrī deposits have been known chiefly for the golden-hued *svarṇa-mākṣika* or *sonāmukhī* or chalcopyrite, the principal copper mineral. The deposit also contains a critical amount of *kāṁsya-mākṣika* or *kāṁsāmukhī* or bronze-looking pyrite which on bacterial oxidation produced ferrous sulphate and ferric sulphate which in turn gave commercially exploitable aluminium and copper sulphates. The other important constituent in the deposit has been *sehtā* or ferro-cobaltite, a cobalt-iron sulpho-arsenide, the topic of which we would like to discuss next.

Sehtā has long been known to the people of Rajasthan as occurring in the copper mines of Singhana, Bagor (2 miles south of Khetrī town) and Babai (8 miles south of Khetrī). The first published notice of this mineral appeared in Capt. A.E. Boileau's 1831 paper on Singhana Copper Works mentioned earlier:

> This *sehtā* looks like a fine grey sand, having the appearance of iron filings mixed with minute particles of silex and mica; its value, at the mines, being from 40 to 100 rupees per maund. It is only produced at Babai, and is used as an ingredient in making glass, *kāñc*, or women's wrist ornaments, *cūrā*, which seem to be made of a vitrified substance, coloured with lac, *lah*, and other pigments.

Brooke reported in 1864 that in Bagor, copper ore was intermixed with cobalt mineral, the latter alternating in thin layers. During Brooke's time (1864), the price of *sehtā* was Rs. 50 for Jaipur *maund̤* of 53 lbs; 'not more than 200 lbs was produced at any particular mine in the year'.

Ball quoted 'Rājpūtānā Gazetteer' stating that '*sehtā* was formerly extensively used for colouring enamels and bangles blue and producing a delicate rose-red colour on gold, an art unknown in Europe'.[43] The process was 'a secret known to certain families' and probably involved roasting *sehtā* and obtaining cobalt oxide which provided different shades of blue. Brooks had written in 1864 that 'pounded fine and exported, this mineral finds its way all over India. It is largely used in enamelling, forming the beautiful blue enamels which natives, proficient in this art, produce'.

Middleton was the first to describe 'the mineral from Jaipur, steel-gray inclining to yellow, probably cubical' and analyse it: cobalt 64.64 per cent, sulphur 35.36, the suggested formula being Co_2S (*Chemical Society Memoirs & Proceedings*, Vol. 3, p. 39; *Philosophical Magazine*, 1846, p. 352). Nicol in his *Manual of Mineralogy* (1849) corrected the formula to be CoS, and named it as Jaipurite (in Punjab it was known as Zaffre or Jaipuri ?).

F.R. Mallet suggested in his classic paper 'On Cobaltite, Danaite and

Jaipurite from Khetrī' (*Records GSI*, Vol. 14, pp. 190-6) that none of the Khetrī samples so far analysed show cobalt in simple sulphide form; the mineral contains arsenic and iron as well.

Mallet purified the mineralogical constituents of *sehtā* with great care, first grinding and liberating the grains and then using Schulze's elutriator and another make-shift elutriator and finally removing iron sulphides by the use of strong magnet. The purified cobaltic part showed 'two different minerals, easily distinguishable under the lens of their crystalline form'. The two kinds were separated with great manual care.

The first kind was noted to be steel-grey, occasionally with a slight reddish tarnish, sp. gr 6.00, its crystalline form mostly cubes, occasionally pentagonal dodecahedron and octohedron. The chemical analysis (percentages): sulphur 19.46, arsenic 43.87, cobalt 28.30, iron 7.83 suggested that it is ferro-cobaltite. Antimony and nickel were detected in traces. Mallet noted a few paler colour samples presumably containing a smaller percentage of iron.

The second kind showed orthorhombic crystals, steel-grey with yellowish tarnish, having the form of arsenopyrite. Mallet suggested, on the basis of very strong reactions for iron, that the crystals are danaite. Mallet finally concluded that *sehtā* contained cobaltite and danaite but no simple cobalt sulphide such as jaipurite.[44]

Reviewing experimental evidences recorded by Mallet as given above, the present author has the following conclusive remarks to offer about the composition of Khetrī *sehtā*. Dana's *Textbook of Mineralogy* (John Wiley 1961 edition) mentions isometric cobaltite CoAsS (35.5 per cent. Co theoretical) and orthorhombic arsenopyrite FeAsS as the extremes of a family of minerals, the intermediate minerals deserving the label of ferro-cobaltite: (Co, Fe) (AsS). In this family of arseno-sulphides, cobalt and iron mutually replace each other. Danaite having 3-9 per cent cobalt naturally resembles arsenopyrite, both being orthorhombic. The Khetrī cobaltite on the other hand is cubic. The present author has computed from Mallet's data of chemical analysis (provided earlier) that the 'first kind' steel-grey cubic material with sp.gr 6.00 corresponds to the formula ($Co_{0.8}Fe_{0.2}AsS$), i.e., it is a ferro-cobaltite with not more than 20 per cent atomic replacement of cobalt by iron. Paler samples have a lower proportion of iron.

There may be some nickel in the lattice. The proportion of high-iron danaite in *sehtā* is small. Thus Khetrī *sehtā* may be stated to be a mixture of ferro-cobaltites which was extracted and marketted for centuries by the people of Rajasthan who did not know before Mallet the composition of the substance. Mallet also found in Nepal 'a specimen of cobaltiferous matt' assaying 11 per cent cobalt which presumably came from the local copper mines.

Sulphates of copper, aluminium and iron

Apart from copper ore and *sehtā*, the other principal products from the sulphidic ore mines of Khetrī were the vitriols or *kasīsa/tutiyā*: *nīla tutiyā* or blue copper sulphate, *harā / pīla kasīsa*, green/yellow ferrous/ferric sulphate, and *mūṅgyā kasīsa* or aluminium sulphate which was converted to alum or *phiṭkarī*. All these water-soluble products were sold and used as mordants for dyeing and in indigenous medicinal compositions. In Kolihan and other mines the various sulphates were prepared by the Hindus, whereas copper mining and smelting were in the hands of Muslims.

The formation of the above-mentioned sulphates in nature owed partly to a biological or bacteriological process, viz., the action of *Thiobacillus ferroxidans* acting on pyrite FeS_2. The chemolithotrophic bacterium derives the necessary energy from the oxidation of ferrous ion and reduced valence inorganic sulphur compounds. Ferric sulphate thus produced convert copper and aluminium into their respective sulphates in an oxidative manner, and get back into the ferrous stage when the cyclic, bacterial and enzymatic process continues.[45] *Thiobacillus thiooxidans* oxidise elemental sulphur and thiosulphate into sulphate and assist in the aforementioned steps of conversion of iron, copper and aluminium minerals to their respective sulphates. The two species of bacteria naturally co-exist in mine water. Whereas Ball mentioned the sale of pyrite or *kāṁsāmukhī* for medicinal purposes in the Calcutta market at Rs. 6-10 per *mauṇḍ*, and 'its alteration by oxidation' in the mine,[46] he and the contemporary scientists were not aware of the bacterial conversion process and the bio-geological phenomenon which had been going on throughout the ages.

In a well-researched paper, T.N. Lung (1986) has documented[47] that the Chinese people were the first to observe the blue colour of the mine water in the copper-iron sulphide ore occurring region, and to produce cementation of copper from the vitriol spring by using iron powder. We now know that the phenomenon involved was a simple displacement chemical reaction.

Cu^{2+} (solution) + Fe (solid) $\rightarrow$ Fe^{2+} (solution) + Cu (solid)

Commercial production of cement copper started in China in AD 1086, but as early as third century AD, a Chinese document mentioned that sour wine (acetic acid) kept in a copper vessel gives a solution from which copper can be precipitated by using iron. *Dhātuvāda*, a ninth-century Indian *Rasaśāstra* text mentions the same recipe (DV. 2-3). James Prinsep (*Journal of Asiatic Society of Bengal*, Vol. 4, October 1835, p. 582) reporting on the Singhana mine specimens, noted:

> There are two very beautiful specimens of *virgin copper* in mammellated concretions, coating oxide of iron, which appear to have been formed by a natural decomposition of the blue soluble sulphate on coming in contact with iron ore.

Prinsep's observation was very significant and the only modification it needs is that stray metallic iron and not 'iron ore' must have been the reactant. We do not know whether Indians ever utilised this reaction in the pre-modern period for obtaining copper metal, as the Chinese had done.

The manufacture of blue vitriol (copper sulphate), alum and green vitriol or copperas (iron sulphate) in Khetrī mines was described and illustrated by Colonel Brooke. (Vide Plate 30) In 1864 there were twenty factories at Khetrī and about double the number at Singhana.

> The broken shale from the mine which contains the salts was placed in earthen *gharā*s, together with the crust from the refuse heaps of previous lixiviations, and water was added.
>
> Each charge of shale was exposed to three changes of water, and the water itself was changed from one *gharā* to another till it had taken up the sulphates from seven different steepings. It was then of a thick dirty-bluish colour, and taken to the boiling house, where it was boiled in earthen *gharā*s; when sufficiently concentrated, it was left to cool, and thin sticks being introduced, the blue vitriol crystallised on them.

> The mother liquor was then poured off and again boiled, and on the addition of saltpetre, the alum crystallised at the bottom of the vessel. The residual sulphates still in solution are allowed to crystallise out by exposing the mixture to the sun.
>
> Seven to eight maunds of shale, with an equal quantity of the refuse, on which the exposure had caused the salts to crystallise out on the surface, yielded four seers of blue vitriol, four seers of alum, and 20 seers of mixed and impure sulphates, including copperas or iron sulphate. The selling prices per maund were for the first Rs. 14, for the second Rs. 4, and for the third Re.1 The Rājā claimed one-sixth of the gross produce as his royalty, and it appeared to Colonel Brooke that the trade was not a very profitable one. Large iron vessels were sometimes used for boiling down the liquor. Traces of nickel and cobalt were found in all these products.[48]

There are numerous accounts of the ancient manufacture of alum in Kutch. The alum pits at Mhurr or Madh had been operative ever since Rao Goharji's time in AD 1716 (*Gleanings in Science*, No. 36, December 1831, pp. 384-5). Towards the end of the eighteenth century and even up to AD 1818, the export of alum amounted in some years to several hundred thousand *maunḍs*, which chiefly went to Gujarat and Bombay to be employed in dyeing. Wynne described the manufacturing operation (*Memoirs GSI*, Vol. XI, p. 88).

The rock containing the materials is a pyritous dark-grey or black shale, which is in close association with a soft aluminous pseudo-breccia of the sub-nummulitic group. This shale was excavated from pits and mines during the rains, and was exposed for four months, 'a slow combustion taking place owing to the decomposition of the pyrites' (actually this was the phenomenon of bacterial leaching explained earlier).

> It was then spread in squares resembling salt pans and sprinkled with water. After about 12 days it consolidated into efflorescing mammillated crystalline plates or crusts called *phiṭkari-kā-bīj* or seed of alum (actually aluminium sulphate). These crusts were boiled in large iron vessels (luted inside with lime), together with saltpetre (or other potash salt), in the proportion of 15 of the 'alum seed' to 6 of the latter; when it had settled, the liquor was placed in small earthen vessels somewhat the shape of

> flower-pots, and crystallisation took place in three days.
>
> The crystals (of real alum, i.e., potassium aluminium sulphate) were again boiled, concentrated, and ladled into *gharā*s with small mouths; these were sunk into the ground to prevent their breaking, and in five days the alum was found crystallised in masses. The vessel was then broken and the alum was stored.[49]

After 1867, the demand for Kutch alum ceased. This was attributed to the State having mismanaged the monopoly, the competition in Bombay with Chinese and English alum, and to the Kutch alum being discoloured and giving a tinge of colour to cloth. Similarly, the earlier manufactories of alum in Sind, Kalabagh in Punjab, Salt Range, Kumaūn district, etc., were closed down on account of inefficiency in production and market competition. Nepal produced aluminium sulphate, an exudation upon soft rocks; this brownish-white *śilājīt*, assaying 95 per cent aluminium sulphate and 3 per cent iron, was sold as a costly medicine, not to be confused with black *śilājīt*, a bituminous substance (*Journal Asiatic Society of Bengal*, Vol. II, 1833, pp. 321, 482 and 605).

The green vitriol or copperas was known to Indians as *kāhī* and *harā kasī*s. We have mentioned earlier, that after crystallisation of blue vitriol and alum, green vitrol or iron sulphate was obtained in the Khetrī Works (1864) as the cheapest residue.

Ball quoted a 1809 record stating that 'many years ago' there had been a manufactory of *kasī*s or copperas at Tilkap, six miles from Rohtāsgarh, on the road to Dehri in Bihar. The source material was the Bijigarh shales having thick efflorescence of sulphate of iron. In 1833, J. Stevenson published an analysis of the local product from Bihar (*Mem. G.S.I.*, Vol. VIII, p. 121): iron sulphate 39 per cent, iron oxide 36 per cent, magnesia 23 per cent. The material was used at that time by the dyers in Patna.[50]

Zinc in medieval India

One of the most spectacular achievements of pre-modern India in the field of metallurgy has been that of zinc. We have discussed the primacy of India in ancient brass and the zinc metallurgy elsewhere.[51] A very startling fact is that the vast information on the subject was unknown to us till in the 1980s several scholars such as Willies, Craddock, Hegde, the present author, etc., unearthed the same through field, laboratory, and literature investigations.

Lynn Willies of the Peak District Mining Museum, England, investigated and reported on three areas of ancient mining in Rajasthan: Rājpura-Daribā, Rāmpura-Aguchā and Zāwar, related to zinc-lead-silver ores.[52, 53] There are evidences that these three metals were extracted from the ores in ancient India. Mining started in Zāwar (near Udaipur) in the fourth century BC (C-14 dating) if not earlier, and went on for centuries.

There was a boost in mining and metallurgy of zinc and lead in Zāwar around thirteenth century AD, and this went on for five centuries. The Hindu rulers of the Udaipur area managed to keep their technology secret. Neither the Muslim nor the British rulers could gain access to the technical information. Thus, Abul Fazl in 1595, Tod in 1829, Brooke (1850) and even Valentine Ball[7] in 1881 could scarcely throw light on the well-guarded secrets of the Zāwar Mines. On Account of famine and Marāṭhā invasions, mining and smelting had stopped around 1812.

The host rock of the Zāwar mines is sheared dolomite. Within the beds, zinc occurs as sphalerite or as marmatite in which ZnS is in solid solution with iron sulphide. Also associated with the rock are galena, hydrozincite, pyrite, silver, etc.

One of the mines investigated by Willies[52] was of commodious proportion — Pratapkhan or Pratap's mine — in which Rāṇā Pratāp Singh, rival of Akbar, took refuge during AD 1595-1600. The mining system had adequate timber work, fire-setting facilities, mining tools, pitchers for manual dewatering, etc.

Manually picked sphalerite-rich particles were roasted and then reduced in brinjal-shaped retorts: the zinc vapour was condensed as liquid under reducing condition and collection. Freestone *et al.* estimated through the analysis of the residue in the spent retorts (about six lakh tons), that nearly one lakh tons of zinc must have been produced during the period of thirteenth-seventeenth centuries.[54] Six thousand tons of zinc metal are still left in the retort residues and recoverable by modern technology.[55]

There are also in this area at least sixteen deposits of glassy slag, totalling 30,000 to 35,000 tons which are residues after recovery of lead and most possibly silver by the cupellation process. It is well-known that Rajasthan lead ores are argentiferous; lead concentrate in the modern Zāwar plant assays nearly 0.1 per cent silver. The Moghul armies wanted to overpower Udaipur territory mainly in order to have monopoly of trade in zinc, brass, lead and silver. (Vide Plates 20-4).

Lead in medieval India

There are extensive old workings of copper, lead and zinc ores in the areas of Ambājī (24°20' N, 72°51' E), Gujarat, and Basantagarh and Deri (24°23' N, 72°50' E) in the neighbouring Sirohā district of Rajasthan. Large dumps of ancient slag, metallurgical tools, smelted metal and potteries have been found at Ambājī and behind the group of Jain temples in adjacent Kumbariā (24°19' N, 72°51' E). The metals were used during the Pre-Christian era Śuṅga Period and the Post-Christian era Kṣatrapa Period. [34,36] Second-fourth century AD lead and silver coins have been recovered from this area.

C-14 dates indicate extraction of copper, zinc, lead (and possibly silver) during AD 1070-1440 from the Ambāmātā deposit in Gujarat and Bandlomattu-Agniguṇḍalā area in the Andhra Pradesh. At Agniguṇḍalā, mining was done[35] in the nature of long, open trenches as well as along the dip of the lodes reaching a depth of 30 m. The ancient Gavulabhavi lead deposit in the Cuḍḍapah district, Andhra Pradesh, was developed similarly through 280 m long surface diggings and underground shafts reaching 50 m below the surface.[56] At Agniguṇḍala, extensive mining, mineral processing (crushing, grinding and beneficiation) and smelting (of copper and lead) took place, which is evident from the rock dumps, slag heaps and appliances such as cupolas, crucibles and tube-like bodies, tapering somewhat and closed at one end. A few pieces of lead were found in the heaps assaying 99.5 per cent lead and 220 ppm silver.[35]

Ball testified to the extent of ancient lead mines in southern India, Afghanistan, Baluchistan and Rajasthan.[57]

Zaṅgamrajupalle lead mine (14°46' 30" N. 78°56'30" E) had been described by King and Newbold. King's account (*Mem. GSI*, Vol. VIII, p. 273) states: "The old and now deserted lead workings are at the south end and east side of the low ridge (near Cuḍḍapah) just north-north east of the village." Further south and west of the village, there are again numerous old galleries excavated in the same series of beds, from which galena ore had been removed. Captain Newbold described (*Journal Royal Asiatic Society*, Vol. VII, 1843, p. 217) having descended a considerable depth into one of a group of upwards 50 mines. Some of the masses of rock exhibited signs of blasting.

The origin of the extensive excavations in this tract is 'lost in obscurity'. The tradition is that these mines were worked by the Hindu kings of Vijayanagara and then the Muslim Nawābs of Cuḍḍapah during the reign of Halim Khan, and later by Hyder Alī and Ṭīpu Sultān. A sample of ore in this area assayed by Mallet in 1879 contained 78 per cent lead and more than 22 oz of silver per ton of lead. Dr. Balfour has been quoted by Ball (Ref. no. 7, p. 232) as saying that the people of Zaṅgamrajupalle in Cuḍḍapah used to 'extract silver from galena by a tedious, clumsy and expensive process in which the lead was all lost'.

In 1846, Captain Newbold described (*Journal Asiatic Society of Bengal*, Vol. XV p. 389) the lead mines in the vicinity of Gazalpalle (15° 26' N, 78' 45' E). A broad deep trench or quarry cut in these rocks is 'supposed to have been the work of the Moghuls'. In the recent times, Rao and Rao have mentioned some of these ancient workings.[56]

Some old mines situated near Jogā on the Narmadā in the district of Hośaṅgābād produced mainly lead ore (*Records GSI*, Vol. XII, p. 174). These were locally known as *Cāndī-kā-Khān* or silver mines. Presumably silver was obtained from galena. Mallet analysed the silver content to be about 21 ounces per ton of lead.

A tract of country on the banks of the Tons river (north-western Uttar Pradesh) partly included in the Sirmur State and partly in Jaunsar district, neighbouring to the site of Swinj (30° 43' N, 77° 47' E) had many ancient lead mines as surveyed by Captain Herbert in 1826 (*Indian Review*, Vol. II, 1826, p. 349). Near Bhatnor or Barthenole (30° 44' N, 77° 46' 30" E) the limestone rock was soft permitting galleries to be made in erect position. At Borela in Jaunsar, however, the passages were narrow and tortuous. The quantity of lead ore which had been removed 'must have been enormous, since at Borela alone there were upwards of 80 galleries'. The mining industry there however dwindled rapidly;

Medlicott who visited the locality in 1862 wrote (*Mem. G.S.I.*, Vol. III, p. 179) that only one mine was open at that time.

Dr. Lord, a medical man reported in 1837 a wonderful lead ore mine at Feringal (34° 45' N, 68° 30' E) in Afghnistan; 'the shafts and galleries showed a most surprising degree of mining knowledge' (*Journal of Asiatic Society of Bengal*, Vol. VII, p. 533). Such mines might have supplied lead to the Moghul India. The lead ore mines of Sekran (27° 48' N, 66° 24' E) in Baluchistan were very extensive but found deserted in 1887. One had to crawl through the tortuous passages for mining the ore. The vast quantities of slag lying about on the surface testified to the former extent of the industry. Some ore was smelted simple open furnaces by the Mardui tribe of the Brahuis.

As we have already told, neither the Muslim nor the British rulers were aware of the details regarding mining and smelting at Zāwar near Udaipur in Rajasthan. The lead ore technology at the Tārāgarh Hill (26° 28' N, 74° 41' 30" E) close to the city of Ajmer has been however well-documented by Captain Dixon[58] and quoted by Ball.[59] Dixon visited the mines in 1830 and wrote :

> That they have been in use for a long time, doubtless many centuries, is satisfactorily attested, by the number that are now lying unemployed.

The ores were of three classes, namely oxides, carbonates, and sulphides; the last variety, galena, was also used as *surmā* for anointing the eyes. The ore, brought to the surface by the miners crawling through the tortuous passages, was crushed, ground and washed by hand in conical wooden tubs. The heavy galena particles settling faster were separated, dried and then 'kneaded up with equal weights of cowdung into lumps of the size of a pigeon's egg'.

> When dried, it was ready for the furnace, a simple structure of clay, 3 feet high, with an upper internal diameter of 11 inches and a lower one of 10 inches. Three clay twyeres were luted into the base and served to communicate the blast from three pairs of bellows. Owing to the presence of cowdung, the quantity of charcoal fuel required was small, only 50 to 75 p.c. to the weight of fused metal. Eact furnace turned out 3-4.5 maunds of lead daily. The ores yielded 31-58 p.c. (average 40 p.c.) lead.

There are records that towards the end of the eighteenth century, the Marāṭhās farmed the Ajmer mines. In 1818, the first British superintendent took over. At that time 10,000 maunds of lead were being produced per year and solve at Rs. 11 per maund. In 1846, the cost of production and the market price of the product (Rs. 16 per maund) went up, and thus Ajmer lead could not compete with English lead. The Ajmer magazine ceased to take the metal and the mines were closed. 'The final closing of the Ajmer mines was due to the Mutiny (1857), and an anxiety on the part of the authorities to make lead for bullets scarce'. That was the inglorious end of the pre-modern lead technology in India!

Gold in pre-modern India

Gold and silver were widely used in the Harappan civilisation. At Hatti, Karnataka, a narrow reef had been worked by the ancients to a length of some 1500 m and to a depth of 190 m by fire-setting alone. One of the C-14 dates there is 760 ± 150 BC. During the pre-Christian era, Hatti and Maski in the Raichūr doab were the main sites for gold-working. Aśoka's Minor Rock Edicts were found on the Maski gold field and the neighbouring diamond fields.

Hatti continued to yield gold in the post-Christian eras, and we have C-14 dates of AD 30 and AD 110 in Oakley's 80 m depth shaft. The first C-14 dates for the famous gold field of Kolar have been obtained by D.P. Agrawal;[33] Champion reef at 50 m depth in Kolar was reached before AD 475. A later date of AD 685 in superficial layer suggests difficulty with the hard rock and retreat mining.

Active mining for reef gold went on for centuries in many gold fields of south India, such as Bellāra, Kabligatti near Pangumpilly in Salem district, Wynāḍ and elsewhere. The Wynāḍ tract constitutes a terrace of mountain-land, intermediate in position between the low country of Malabar and the lofty plateau of the Nīlgiri mountains. Pliny mentioned gold of this area extracted by 'Nareae' or the Nairs of Malabar. Ball quoted[60] Brough Smith's 1879-80 report on the Wynāḍ gold-fields. In the tract of 500 square miles surveyed, there were quarries on the outcrops of veins, vertical shafts, adits, vertical shafts with adits and shafts on underlie.

> The most remarkable were the vertical shafts, even in solid quartz, sometimes 70 ft deep, with smooth and quite plumb sides.... Stones were pounded with hand-mullers, and the pounded stone washed in a wooden

dish and treated with mercury.[60]

Gold from alluvial washings

For centuries, the Indian rulers hoarded gold, a large fraction of this, specially from the Roman Period onwards, being through imports. Rock mining could never meet the entire imperial requirements. The other indigenous source has been the placer (alluvial or detrital) deposits in the auriferous sands and gravels of a large number of Indian rivers which pass over auriferous rocks.

Tabling action or flowing-film concentration has been a natural phenomenon taking place along the bend of the rivers and during the recession of water after the rise during the floods. Heavy minerals such as gold have been concentrated on the river banks for ages, and toiling masses have further purified the same through painstaking efforts. This kind of laborious work by a large number of poor people, spending long hours to produce a small quantity of precious materials for the affluent, has been the feature of the ancient world including India. Ball and others[61] have compiled bibliographies on pre-modern gold washing in India. We would cite a few specific examples.

Gold dust was known to occur in the upper Godāvarī region, specially near Bhadrachalam (17° 40' 30" N, 80° 51'45"E) and Mariguram (18° 17' N, 80° 36' 30" E), where the tributaries overflowing gold-bearing rocks joined Godāvarī. The people who washed gold were known as *sonjharis* or gold seekers.

In the nineteenth century, the census returns showed many professional gold-washers operating near the Kasai river in Midnāpur, West Bengal, and near the Suvarṇarekhā (means gold-lined) river in the Singhbhūm area in Bihar. The wooden dish used by the *jhoras* (washers) measured about 28 by 18 inches, hollowed somewhat eccentrically to a maximum depth of 2½ inches. A scraper formed of a flattened iron hook, set in a handle, was used to collect the auriferous sand and gravel which accumulated in the angles formed by the rocks in the bed of the stream. The dish containing the particulate matter was washed in shallow water, employing manually generated vibratory motion. Separated from coarse pebbles and fine-light gangue, the final product used to be 'a residue of black iron sand in which the specks of gold were readily apparent'. (Vide Plates 3-4)

On both sides of the Ebe or Ib river in Orissa, there were tracts, honey-combed, with shafts sunk by successive generations of gold-seekers. Some of the shafts were 10-30 feet deep. The principal rivers of this tract are the Ib and the Mand which carry alluvial gold. In the former States of Gaṅgpur, Jaśpur and Udepar (latitutes 21-25° N, longitutdes 83-84° E), rock mining as well as gold-washing used to be done.

Close to Sambalpur town in Orissa (21° 28' N, 84° 2' E), the banks and bed of the river contain gold. The various tributaries of the Mahānadī which traverse the metamorphic rocks are all probably more or less gold-bearing. Diamond particles are also available in the river sand (hence the name *Hīrā-kuṇḍ*), but these come from the lower Vindhyan rocks.

Major J.R. Ouseley described the indigenous 'process of washing for the gold dust and diamonds at Hīrākuṇḍ' (*Journal of Asiatic Society of Bengal*, Vol. VIII, 1839, pp. 1057-8), operated by gold-washers of Jhunan who used to travel to different localities when the water fell after the rains. Ouseley suggested:

> Gold washings might be undertaken (at Hīrākuṇḍ) on mechanical principles which would, by reducing the manual labour, make the speculation highly profitable in gold dust alone, setting aside the diamonds.

He even elaborated the construction of such a possible machine featuring the principles of sieving, gravity settling and tabling.

Captain Jenkins forwarded a lengthy paper on the production of gold in Upper Assam, written by one Muniram and this was published in the *Journal of Asiatic Society of Bengal* (Vol. VII, 1838, pp. 621-7). In that paper we are told that before the British took possession of the country, gold-washing had been practised extensively in Assam. The *Sonwals* paid a yearly tribute of 4000 *tolā*s of gold (Rs. 16 a *tolā*), i.e., Rs. 64,000. This appears plausible from the published data of Moghul-Ahom trade during the first half of the eighteenth century, when a substantial amount of gold used to be imported to Moghul India from Assam.[62]

Muniram referred to several methods of collecting gold as practised in pre-modern Assam. The *kachari*s or *cassari*s used to wait for a rise in the river and immediately afterwards, on its fall, scraped up the newly deposited sand, which they washed for gold. Other Sonwals washed for gold during the dry season. The gold-washers in the Sitang area got the gold by washing the moss and slime which were scraped from the rocks in the bed of the river.

The final washing used to be done in long wooden troughs placed at a small angle, allowing flowing water and dirt to run off and allowing gold and heavy fraction to settle in. The settled dust was made to react with mercury, when gold formed into a lump amalgam. The lump was put into a shell and on a fire of wood charcoal, when mercury evaporated. The residual gold was further purified by salt roasting. While publishing Muniram's 1838 report, the editor of the *JASB* commented (*JASB*, 1838, p. 627):

> Might not the galvanic magnet be advantageously employed in freeing the washed sand of the ferruginous particles? We (in Geological Survey of India) have frequently employed the common magnet in the examination of small specimens of these sands with advantage. The use of mercury might thus be avoided.

Gold dust was available in all the tributaries joining the Brahmaputra. The banks of the Dihong and the Jugla were particularly rich. The Ningthee river sand in Manipur State also yielded substantial quantity of gold dust (*JASB*, Vol I, 1832, pp. 148-9). The process of gold dust recovery from the river sands of Gomatī in the sub-Himālayan terrain and of Beas or Vipāśā in Punjab (*JASB*, 1835, pp. 279-82; 1847, pp. 266-8) was quite similar to the technique followed in Assam. The British authors commented about the inefficiency of the gravity separation and amalgamation/mercury-recovery steps. Since gold dust settled with 'very heavy' black ferruginous matter in the washing trough, there was adequate scope of trying magnetic separation method. There was also the proposal for mechanisation; operation at 'a larger scale with machinery turned by the river current, might find it pay handsomely, but only by personal supervision' (*JASB* 1847 paper on Beas sand, p. 268).

Abul Fazl neatly summarised the position of gold resoures in medieval India:

> Although gold is imported into Hindustan, it is to be found in abundance in the northern mountains of the country, as also in Tibet. Gold may also be obtained by the *salonī* process from the sands of the Ganges and Indus, and several other rivers, as most of the waters of this country are mixed with gold; however, the labour and expense greatly exceed the profit.[63]

Ball reported the continuing supply of gold from Ladakh and Tibet to India. From the ancient eras there had been a perpetuating myth about 'gold-digging ants'. This mystery was clarified with the knowledge that the miners of Ladakh and Tibet preferred to work during winter in the mines under the frozen soil. Their tents set above the series of pits were made of black felt-like material, manufactured from the hair of the Yak. Digging was done by horn of the wild sheep tipped with iron mounted on a wooden handle. The miners, thickly clad with furs, looked like giant-sized gold digging ants; they used animal skins and hairs to arrest gold particles flowing with the stream.[64]

From time immemorial, India imported gold and silver from whatever countries convenient — Tibet, Burma, the European countries, Latin America through Europe, etc. India has been the veritable sink of precious metals with huge export list and incommensurate import on account of the poor people having little or no purchasing power. The rulers preferred to have precious metals in return of the export commodities. For gold-washing, 'the labour and expense greatly exceeded the profit' as Abul Fazl put it, and yet for the poor and exploited people this was often the bare means for survival.

The crude gold extracted and alloyed gold (and silver) coins had to be purified and assayed. This necessitated suitable exercises in royal *kārkhānā*s which had smelting, assaying and minting facilities. Ṭhakkura Pherū outlined the technique for assaying gold and silver coins by melting and using touchstones.[4, 65] These techniques were in vogue for centuries and underwent some improvement.

Abul Fazl explained[5] how standard alloys of gold, with varying percentages of gold in silver, were prepared and their streak properties on touchstone served as calibration indices for the approximate estimation of gold in an unknown sample.

The *salonī* process of assay started with impure gold made into thin plates, a few were weighed and interspersed with saltpetre and brickdust and put to fire repeatedly. The final product was pure gold, which was weighed, and ash (*salonī*) containing all silver and some gold. Recovery of the gold lost in ash was possible by amalgamation procedure; smelting with lead gave recovery of silver. The *salonī* process of assaying impure gold was followed for centuries, even by the Calcutta goldsmiths as late as 1827. A. Campbell reported in 1834 that the same method was being followed by the Nepalese to assay the impure Tibetan gold which they were importing at the price of Rs. 15 per *tolā*. The use of saltpetre or salt was meant to convert silver into a salt and remove it as ash or *salonī* (*JASB*, December 1834, pp. 622-8).

Another method cited by Abul Fazl[5] for removal of silver from gold was by smelting the alloy first with copper and then with sulphur, when silver part would go to the surface as ash. Silver could be extracted from this ash by re-smelting with borax, natron and lead metal. The small imports to India of lead and mercury were chiefly on account of the necessities of purification and assay of gold and silver samples. Abul Fazl quoted[5] Al-Berunī in providing the specific gravity values for gold, silver, copper, other metals, minerals and gems.

Building stones

The systematic use of building stones in India probably started during the Indo-Greek period, or even earlier during the contacts with the Persians as claimed by Wheeler and Piggott.[66] We find that during the Pre-Christian era periods, many kinds of hard stones were used at Taxila:

Saṅg-ī-abrī (a lower cretaceous shell-limestone of a variegated chocolate and yellow colour, abundant in the Hazārā and Attock districts), granite, basalt and several kinds of sandstone such as locally available grey *tarakī*, spotted red and white variety from Mathurā, buff-coloured fine-grained hard variety from Cunār near Vārāṇasī, etc.

The traditions related to the stones continued right across the medieval ages. V. Ball summarised the pre-modern era scenario from the standpoint of a geologist, categorising the discussion under sandstone, granite, basalt, marble, limestone, laterite, etc.[67-69]

The red sandstones of Cunār, Mirzāpur, etc., have been used ever since the period of Aśoka who built with these materials the great monoliths or *lāṭ*s bearing his third-century BC edicts. These were generally polished throughout the portion intended to be exposed, and surmounted by carved and ornamented capitals, upon which figures of lions or elephants were placed. Ball tabulated the various monuments and structures of pre-modern India built of this sandstone (Ref. no. 67, p. 115), and commented that

> the ancient pillars of Vindhyan sandstone have been instrumental in annihilating time by preserving, in an imperishable record, fragments of the history of upwards of two thousand years ago.

The traditional continuity is still visible. As Ball pointed out, the *ghāṭ*s at Vārāṇasī, the palaces, walls, minarets, temples, etc., are built of Cunār sandstone. Similarly, sandstone of different hue and colour were used in buildings like palaces at Jaisalmer, Bikāner, Bharatpur, Fatehpur Sikri, Jumma Masjid at Delhi, etc.

The Vindhyan sandstone beds often spread for long distances *without* any joints or fissures to break the continuity, in consequence of which very large blocks can and have been extracted for various purposes. Mallet mentioned large blocks found near the quarry of Rupas: 3' diameter 34' long circular column and a 4' × 5' × 42' 6" long parallelopiped weighing nearly 60 tons. The material probably belonged to the Aśokan period. Ball noticed and reported indigenous devices for lifting and transporting large masses of stone on very solid wooden wheels (Ref. No. 69, facing p. 544). (Vide Plate 5)

Granite was employed in the construction of many temples in Bihar, some of the Buddhistical rails at Gayā, etc. On Mahendragiri hill in the district of Gañjam, huge blocks of porphyritic gneiss, one measuring 9'×4'×3', have been found. These were brought during some ancient period for the construction of a temple (unfinished). A polished slab of quartzo-felspathic gneiss in the Durbār Hall at the Tanjore Palace measure 18' ×16' × 2'. Some of the Mahābalipuram temples show 'an incredible amount of labour out of solid bosses of granite *in situ*'. Quarrying of granite in south India and its export has become a thriving industrial activity in modern India.

Between Gwalior and Goa, Bombay and Amarkaṇṭak, we have the vast accumulation of eruptive basalt rocks known as the Deccan trap. A variety of basalt suitable for construction is 'bluish green, heavy and hard, rings like a metal on being struck'. An ancient form of architecture (earliest 200 BC) consisted in hollowing out and sculpturing the rock *in situ* the temples and dwelling places, of which we have magnificent examples in the caves of Ajantā, Ellorā and Elephaṇṭā.

Some of the Buddhistical rails at Gayā are made of basalt. Temples and images in the ruined city of Gaur and many Burdwān district Hindu temples have been made of basaltic trap from the Rājmahal Hills. Polished specimens have often been confused as 'black marble'.

Marble in India has been well-known for its spectacular use in ornamental architecture. The Dilwārā temples in Mount Ābu, the great Tājmahal, the tomb of Sheikh Salīm Cistī at Fatehpur Sikri (AD 1581) are some of the Indian 'poetries in marble' made out of the rocks in Rajasthan.

The most extensive quarries in Rajasthan correspond to those of Makrānā bordering erstwhile Jodhpur and Jaipur States. The pure white vertical beds are often couple of feet thick yielding reasonably large blocks. These quarries supplied the marble for the famous Tājmahal.

The stone near Jhirri at Alwar is harder to cut than the Jodhpur marble, but is more cheaply raised. The two-mile long quarry provided many monoliths and pillars in the *āmkhās* of the Alwar palace. The Alwar State also provided in the past the black marble of Māṇḍlā and Rāmgarh, and the pink marble of Baldeogarh. Baden Powell reported from Punjab yellow marble of Manairi, grey marble of Bhunsi and the Kashmiri black marble.

The white marble from Raialo or Raiwālā in the erstwhile Jaipur State has been widely used for making *jālīs* or filagree screens. Some of the screens are 'so minutely pierced that they actually look like lace at a little distance'. "Such pure patient workmanship, so dignified and yet so various, is hardly to be found anywhere in the world" (Ref. no. 68, pp. 463-4). This kind of material and workmanship are to be found in the ancient art-works of the Udaipur area and Gujarat also.

The marble of Jabalpur has not been widely used for the constructional purposes. *Kaṅkar* or concretionary carbonate of lime has been used for building purposes. The widespread deposits of limestone in Assam, Sylhet, Bihar and the western India were known in pre-modern India and used partly for construction and partly for calcination to produce lime.

The term laterite has been applied generically to a group of ferruginous rocks predominant in the superficial geology of India. Dr. Buchanan conferred the title 'laterite' (from *later*, a brick) on account of its reddish-brown appearance. The differences in composition and structure make infinite varieties of laterite. Some varieties as those used in the Tanjore Fort Complex are of poor quality, crumbling away under the influences of weather. But other varieties actually harden on exposure and have been widely used in the Rājmahal Hills, Midnāpur, Orissa and the coastal belts, as temples, tanks or bowries. Ball noticed an indigenous procedure of cutting 4-5 ft long laterite slabs. Such blocks were safely used for building works and eventually retained their sharp edges.[67,69]

How could some Indian temples and structures of antiquity withstand the ravages of time and weather, and some others could not? This point needs to be investigated scientifically. Raymahashay and Sharma have recently shown that the rapid decay of the famous Sun Temple at Koṇārk, Orissa, is mainly due to the use of a highly weatherable *khoṇḍalite* rock as the construction material.[70] D.N. Wadia, the famous Indian geologist, studied *khoṇḍalite*, a quartz-garnet gneiss with graphite, sillimanite and feldspar as common associates, available in Orissa, and named it after the Khoṇḍ tribe inhabiting the area. Khoṇḍalite samples obtained from the Koṇārk complex and a quarry in the Dhavalgiri hill near Bhubaneśvar, contains red garnet of the almandine variety which has been found under SEM photography to undergo oxy-alteration into powdery kaolinite and goethite veins. This phenomenon has resulted in greater porosity and reduced mechanical strength of the Koṇārk Sun Temple.[70] It is feared that the Tājmahal marble is likely to be chemically eroded by sulphur dioxide and carbon dioxide present in the exit gases of the neighbouring industries.

Ball referred to occasional uses of serpentinous rocks, soapstone, gypsum and alabaster for constructional purposes in the pre-modern India. Ornamental stones have found limited uses in ancient buildings. There are inlaid works in the Tājmahal at Āgrā for example, made of lapis lazuli, jasper, heliotrope, agate, chalcedony, carnelian, sard, plasma, jade, etc.

Iron ores in Indian antiquity

Only a few remarks can be made about the exploitation of iron ores in pre-modern India, the subjects of smelting, production of pig iron and steel being exhaustive and needing treatment in a separate chapter.

Iron ores have been ever abundant in India, and it is conceivable that when Iron age started around 1200-1000 BC — simultaneously in the northern, southern and eastern parts of the sub-continent — local ores were used. The Gwalior iron ore belt continued to serve Indian iron industries, starting from the Atrañjikhera Pre-PGW era to the Ujjainī Mahājanapada era, then Iron Pillar famed Gupta age and even up to the period of Akbar.

During the Sultanate period, the mining areas lay scattered in the hilly region beginning with Gwalior and extending to the tip of south India. The Cutch iron ore probably resulted in the famous swords of Korij; the Geniza records show (eleventh and twelfth centuries) that the Deccan exported iron and steel to the Middle East. The exceptionally high-grade ores were used to produce damascened steel which had a world-wide reputation. The Gwalior mine was used to produce 'numerous articles' which were sent to the principal cities of the Moghul empire.[71]

From AD 1660 onwards, the Dutch assisted in the exploitation of the Godāvarī belt iron ore, production of metal and export of the same through Masūlipaṭam (Ref.no. 6, p. 192). During the Moghul period the autonomous Ahom (Assamese) dynasties exploited clayey iron ores, their mines often reaching 24 ft deep. The iron manufacturers in the Jaintiā and other parts of the Khāsī hills had a big market in medieval India. Along with bell-metal vessels, iron hoes used to be exported from the Ahom states to the Moghul Bengal (Ref.no. 3, pp. 491-4).

The above informations are indeed scattered and inadequate to cover the time period of one millennium and the vast area of the sub-continent. We obtain some worthwhile glimpses about the indigenous traditions on iron ore processing in India from the British travellers and geologists who left beautiful accounts of their observations made during the eighteenth and nineteenth centuries.[72-85]

Valentine Ball wrote the first comprehensive review on Indian iron ores.[86] Although his lengthy essay came under 'economic geology', oriented towards the possibilities of *future* economic exploitation, there were enough historical overviews outlining the *pre-modern* mining and ferrous extractive metallurgy in the eighteenth/nineteenth centuries. Certain salient facts regarding the iron-ore varieties in pre-modern India are collated in Table 3.4. These show widespread mining of iron ores of diverse types in the sub-continent. Various kinds of technology and furnace were employed to extract iron and produce steel, but this topic is outside the purview of the present chapter.

In the sixteenth and seventeenth centuries, Assam was famous for the manufacture of big iron guns. There were 3000 smelters and smiths in upper Assam. Clay ironstone and limonite have been the principal raw materials in the Brahmaputra valley. In the Khāsī and Jaintiā hills on the other hand, the minute crystals of titaniferous magnetic oxide were 'raked into a small stream of water' (hydraulic mining) and then further beneficiated in wooden troughs. (Vide Plate 9)

In the Pālamau sub-division of Lohārḍāgā district, three different kinds of ore occurred: (a) magnetite, (b) siderite and hematite, and (c) red and brown hematite in lateritic rock. The Agāriā iron-smelters preferred to use magnetite altered in hornblendic or granitic rocks (*Mem. GSI*, Vol. XV, Part I, p. 112). Sambalpur smelters also chose magnetite in metamorphic rock to be reduced by *śāl*-wood charcoal.

Major Franklin examined in 1829 the ancient mines and ore deposits of Jabalpur, Baragāon, Pannā, Katola and Sāgar in central India.[81] The smelters used charcoal, made from specific trees such as teak, *mowa* or bamboo. At Tendukhera, high grade, slightly calcareous ore was used to make *kaccā* iron as well as *pakkā* steel.

The information on the ancient mines around Gwalior was compiled by Ball (Ref. No. 7, p. 394). The rich deposits of iron ore had been formerly extensively worked, but when Hacket examined them (*Rec. GSI*, Vol. III, p. 42) they were 'more or less deserted in consequence of the distances to which the ore had to be taken to the furnaces situated near the fuel'. At that time the nearest forest, from which charcoal could be prepared, was situated 50 miles to the north-west of Gwalior.

Hacket examined the thin laminae of iron ores in the Pār Hill (lat. 26° 2' 30", long. 78° 5') variegated clays which had been extensively burrowed by the miners. Mangor, situated 3 miles north north-east of the Pār Hill produced ore of similar character; the richest seams had been worked by shafts 30-40 ft deep. The ores at Santow (lat. 26° 6', long. 78° 10') belonged to the same horizon as those at Pār and Mangor. The richest seams had been reached by shafts from 50-60 ft deep, from which small galleries were extended. There were smaller mines at intermediate localities, and the ore beds had a wide extension. Von Schwartz reported that iron ores occurred 'in very great quantities' not only in Santow but also in the neighbouring villages of Maesora, Gokalpur, Dharoli, Bamaori, Sipri, Turwaya, etc.

Table 3.4
Indigenous varieties of iron ores in pre-modern India

Location	*Type of Ore*	*References*
Assam (Upper)	Clay ironstone	Hannay in *JASB*, 25, 330
Khāsī and Jaintiā Hills	Titaniferous magnetic oxide	Yule.[80,82]
Birbhūm in West Bengal	Partly earthy, partly magnetic	Hamilton[73]
Bālāsore, Orissa	Lateritic ore	Hamilton[73]
Pālamau, Bihar	Three types of ore, Agarias preferring the magnetic type	*Mem. GSI*, Vol. XV, part 1 p. 112 Ref. No. 7 pp. 376-81
Sambalpur	Magnetite in metamorphic rock	*Rec. GSI*, Vol. X, p. 182,
Tendukhera, Madhya Pradesh	High grade, slightly calcareous	Franklin[81] Ref. 7, p. 386 Ref. 28, pp. 269-94
Gwalior	Rich deposits, now deserted	Hacket, *Records GSI*, Vol. III, p. 42 Ref 7, p. 394
Nimār and Mālwā in Indore Territory	Barwai deposit is chiefly brown hematite	Ref. 7, p. 397
Kāṭhiāwār and Cutch	Laterite of sub-nummulitic group	Ref. 7, pp. 400-2
Hyderābād	Two kinds of ores — lateritic and magnetic sand	Voysey[79]

Location	*Type of Ore*	*References*
Cuḍḍapah and Kurnool	Siliceous hematite and also lateritic	1795 report of Heyne[75]
Mysore	Black sand iron ore in Veṅkaṭa-giri	Buchanan[78]
Cītaldurg in Salem	Non-magnetic ochrous limonite	Heyne[75]
Mālābār or Kerala	Black magnetic sand	Buchanan[78]

Indore territory (Nimār and Mālwā) had long borne a reputation for producing iron, and in the *Āin-ī-Akbarī*, the steel mines of Indore are alluded to, together with those of Nirmal in Hyderabad. The principal localities proceeding down the Narmadā valley, where mines existed, were Chandgarh, Ponassa, Barwai, Kandicote, Bāgh, etc. Within a radius of 4-5 miles around Barwai, there were old mines near the villages of Nāndiā, Koruṇḍiā, Choṭi Modri, Karampura, etc. The ore there is chiefly brown hematite or pure red hematite. Barwai is located 39 miles by rail from Khaṇḍwā and 47 miles from Indore.

Benjamin Heyne,[75] Francis Buchanan,[78] H.W. Voysey[79] and others surveyed and described the different kinds of iron ore being processed in south India during the beginning of the nineteenth century. Buchanan coined the term 'laterite' from *lateritis* (brick) to denote the earthy/clayey iron-bearing red rock. Apart from lateritic rock, black magnetic sand, non-magnetic ochrous limonite, siliceous hematite, etc., were mined and processed for extraction of iron.

The steel mines of Nirmal in the Nizām's territory of Hyderābād, where *wootz* used to be produced, were mentioned in the *Āin-ī-Akbarī*. Malcolmson described (*Transaction Geological Society*, London, New Series, Vol. V, p. 546) the kind of ore processed in Dimdurti. The iron ore grains were diffused in a sandstone-like gneiss or mica schist passing into a hornblende slate. These were excavated with crow-bars, crushed between stones and then beneficiated by washing. Mirtapalli ore was magnetic sand, whereas the Koṇḍāpur raw material was clayey, probably lateritic. Voysey described[79] the making of wootz at Konasamudram, 12 miles south of Godāvarī and 25 miles from Nirmal. When he mentioned that 'two kinds of iron were put in the crucible and no charcoal', probably he meant that wrought iron and high carbon cast iron, prepared separately from two kinds of ore (from Mirtapalli and Koṇḍāpur) were fused together. Both Voysey and the Persian trader, quoted by Voysey, wrongly presumed that the superb quality of wootz was due to the nature of the ore; they did not realise that adjustment of the intermediate carbon content (between those of wrought and cast iron) was the crux in making steel. Of course, the Indian smelters were also unaware of the underlying chemistry in the wootz technology developed by them.

Concluding remarks

In this chapter, dealing with the mining and processing of non-gem minerals in pre-modern India, we have recorded various traditions and achievements in the sectors of: marine salt, riverine bed alkaline salt or *reh*, saltpetre, copper ore, *sehtā* or ferro-cobaltite, sulphates of Cu, Al and Fe known as various kinds of *tūtiyā* or *kasīsa*, ores of zinc, lead, gold, silver and iron. We have discussed the gem minerals in pre-modern India in a separate chapter.[2]

Bulk of the categorised informations has been obtained from the account of travellers and European geologists. The latter have discovered novel features of physical geology in India, invented some specific terms such as 'laterite' and observed new minerals such as seypoorite or jaipurite which is actually a ferro-cobaltite. The famous 'dog metal' or Kyanite mineral having high density, hardness and thermal refractoriness was discovered by Dr. L.L. Fermor (of the Geological Survey of India) in Kharswan of the Singhbhūm district, Bihar.

Through centuries, several unit operations were practised, e.g., evaporation of aqueous solutions, fractional crystallisation, pre-modern and crude techniques of mining, crushing, grinding, rudimentary steps of gravity separation, carbo-thermic reduction smelting, using charcoal as the reductant, amalgamation technique for purification of noble metal, etc.

The status of the mining techniques was not only pre-modern but also stagnant. There was hardly any improvement across the centuries. The Indian society was bereft of any tradition for mechanisation; the mines had to be abandoned as soon as the water level was reached, since no mechanical pump had been invented. There were a few waterfall-driven crushing machines in the Kumaon Hills and Chinese type blowing machines to boost smelter devices in Assam during the nineteenth century, but these were rare exceptions.

Commenting on the lead mines of Ajmer, Captain C.G. Dixon wrote in 1831:

> The system of restricting the communication to the tortuous course of the vein, precludes the practicability of removing the water; in as much as shafts, galleries, adits, and the usual concomitants found in similar cases in Europe, if not quite unknown, are altogether neglected. . . . While we wonder at the simplicity, the leading characteristic of their operations, we are compelled to acknowledge our surprise, that the result are so favourable.[58]

The reputed historian Tapan Raychaudhuri could not defend the lacklustre achievement of Moghul India 'which had little curiosity about the laws of nature'; 'neither the elite nor the mass of producers manifested any curiosity, utilitarian or otherwise about things mechanical' (Ref. no. 3, p. 307).

> The marvels of Mughal architecture were achieved without the aid of wheelbarrows. . . . Ovington ascribed the non-adoption of printing in India to a concern for the scribes' prospects of employment. The Dutch technique of quadrupling the rate of production of iron nails and cannon balls was banned for it might deprive many blacksmiths of their livelihood.
>
> — Ref. no. 87, pp. 18-20

Apart from the oppressive nature of the control of the mines and lack of administrative support, the industrial workers faced insulation from the intellectual world. Whatever scientific curiosity was nurtured by the authors of the *Rasaśāstra* texts did not permeate down to the level of miners on account of the lack of educational and printing facilities.

Dharampal suggested that 'the link between the practitioners of various professions and the professors had largely snapped in India by the end of the eighteenth century', and that 'in a changed political climate such links would have been re-established' (Ref. no. 28, p. 43). We consider this to be somewhat wishful thinking. The aforementioned 'links' had rarely existed. The scholars hardly interacted with the workers, whereas the latter jealously guarded the secrets of their trades. This explains the severe paucity of indigenous literature on the subject. Most of our information has been derived from the writings of the foreign travellers who closely examined the scenario.

There was a certain element of inevitability of the Western people intruding into the Indian sub-continent, first through trade and then through modern technology, initially as traders and then as exploiters. The redeeming feature in our sub-continent has been the gallant and persistent craftmanship of workers in the area of gems and non-gem minerals — in terms of both mining and processing.

References

1. A.K. Biswas, *Minerals and Metals in Ancient India*, Two Volumes, Report to the Indian National Science Academy (INSA), New Delhi, 1991. Vol. I Archaeological Evidences, II. Literary Evidences. Published by D.K. Printworld (P) Ltd., 1996.
2. A.K. Biswas, *Gem-Minerals in Pre-Modern India*, chapter II in the Report to INSA on a project initiated in 1991 and submitted in 1994. This book, chapter II.
3. Tapan Raychaudhuri and Irfan Habib (ed.), *The Cambridge Economic History of India*, Vol. I: AD 1200-1750, Cambridge University Press, 1982, Reprint Orient Longman Limited, New Delhi, 1993. Article written by Irfan Habib on Non-Agricultural Production and Urban Economy Under the Sultanate, pp. 76-7.
4. Bhamarlal Nahta (ed.), *Dravyaparikkha aur Dhatu-utpatti* by Ṭhakkura Pherū, Vaishali, 1976.
5. H. Blochmann (tr. and ed.), *Āin-ī-Akbarī* by Abul Fazl, Vol. I, Reprint by Oriental Books Reprint Corporation, New Delhi, 1977, pp. 15-16, 40-5, 58 and 525.
6. W.H. Moreland, *From Akbar to Aurangzeb*, Oriental Books Reprint Corporation, New Delhi, 1972, pp. 45-8, 100, 118-22, 192-4 and 292.
7. Valentine Ball, *A Manual of the Geology of India, Part III, Economic Geology*, Geological Survey of India, 1881.
8. A.G.F., Eliot James, *Indian Industries*, W.H. Allen & Co., London 1880, Section on Salt, pp. 228-36.
9. Ref. no. 7, p. 488 foot-note.
10. Ref. no. 5, p. 525 also footnote.
11. T.H.D. La Touche, and W.A.K. Christie, The Geology of the Lonar Lake with a Note on Lonar Soda Deposit, *Records of the Geological Survey of India*, Vol. 41, 1912, pp. 266-85.
12. J.B. Auden, B.C. Gupta, P.C. Roy, and Mehdi Hussain, Report on *Reh* Soils, *Records of the Geological Survey of India*, Vol. 77, 1942, Professional Paper No. 1, pp. 1-45.
13. C.J. Brown, and A.K. Dey, *India's Mineral Wealth*, Oxford University Press, Bombay, 1955.

14. Ref. No. 7, pp. 493-4.
15. Ref. No. 7, pp. 498-9.
16. F.R. Mallet, *A Manual of the Geology of India, Part IV, Mineralogy*, Geological Survey of India, Calcutta, 1887, pp. 134-7.
17. George Sarton, *Introduction to the History of Science,* Vol. II, Carnegie Institution of Washington, 1931, Reprinted 1962, Part I, p. 29 and Part II, pp. 766-7 and 1036-40.
18. P. Ray, *History of Chemistry in Ancient and Medieval India*, Indian Chemical Society, Calcutta, 1956. A summary of P.K. Gode's review on early use of gunpowder and saltpetre in India is available in pages 225-9. Also *vide* Iqtidar Alam Khan, Origin and Development of Gunpowder Technology in India, 1250-500, *Indian Historical Review* Vol. 4, no. 1, July 1977, pp. 20-9.
19. Ref. No. 5, p. 58.
20. Ref. No. 6, pp. 18-19.
21. *Ibid.*, p. 215.
22. *Ibid.*, p. 122.
23. S. Chaudhuri, *Trade and Commercial Organisation in Bengal 1650-1720*, Calcutta, 1975 pp. 161-2.
24. Ref. No. 6, p. 320.
25. A.J. Qaisar, *The Indian Response to European Technology and Culture 1498-1707*, Oxford University Press, 1982, pp. 55-6.
26. *Ibid.*, pp. 82-3.
27. Ref. No. 7, pp, 500-1.
28. Dharampal, *Indian Science and Technology in the Eighteenth Century*, Biblia Impex Pvt. Ltd., New Delhi and Academy of Gandhian Studies, Hyderabad, 1983 edn., p. 37 and Robert Barker's paper 'The Process of Making Ice in the East Indies' pp. 219-22.
29. Ref. No. 1, Vol. I, chapter XVII.
30. Lynn Willies, Ancient Zinc-Lead-Silver Mining in Rajasthan, India, *Bulletin of the Peak District Mines Historical Society Ltd.*, Vol. 10, No. 2. Winter 1987, pp. 81-123; also Ancient Lead-Zinc-Silver Mines in Rajasthan, *Mining Magazine*, January 1989, pp. 31-5.
31. D.K. Chakrabarti, Mining in Ancient and Medieval India, in *The Cultural Heritage of India*, Vol. VI, Ramakrishna Mission Institute of Culture, Calcutta, 1986, pp. 188-96.
32. H.V. Paliwal, L.K. Gurjar, and P.T. Craddock, Zinc and Brass in Ancient India, *Canadian Mining and Metallurgical Bulletin*, Vol. 79, No. 885, January 1986, pp. 75-9.
33. D.P. Agrawal, and C. Margabandhu, Ancient Gold Workings: Some New C-14 Dates, *Puratattva, 8*, 1975-6, pp. 138-9.
34. D.P. Agrawal, C. Margabandhu, and N.C. Shekar, Ancient Copper Workings: Some New C-14 Dates, *Indian Journal of History of Science*, Vol. 11, No. 2, 1976, pp. 133-6.
35. Mohd. Ziauddin, Ancient Copper Mining and Metallurgy near Agnigundala, Andhra

State, *Indian Minerals*, Vol. 15, 1961, pp. 117-20.

36. N.C. Shekar, Antiquity of Mining and Metallurgical Activities at Ambaji, Kambaria and Deri, Gujarat and Rajasthan, *Indian Journal of History of Science*, Vol. 18, 2, 1983, pp. 176-83.
37. Ref. No. 3, pp. 369-71.
38. *Ibid.*, p. 491.
39. Ref. No. 7, pp. 239-80.
40. Ref. No. 7, pp. 260-1.
41. V.K.S. Varadan, and M. Narasimhan, Report on Madhan-Kudhan Section, Khetri, Unpublished Report, Indian Bureau of Mines, 1965.
42. K.R. Raghu Nandan, Ancient Mining in Khetri Copper Belt, Rajasthan, India, *in* Scientific Hertiage of India — Proceedings of a National Seminar — ed. by B.V. Subbarayappa and S.R.N. Murthy, The Mythic Society, Bangalore, 1988, pp. 177-82. *Vide* also Bulletin *GSI*, Series A, Economic Geology, No. 47, 1981.
43. Ref. No. 7, pp. 324-5.
44. Ref. No. 16, pp. 16-18, 22 & 27
45. L.E. Murr, A.E. Torma, and J.A. Brierley (ed.), *Metallurgical Applications of Bacterial Leaching and Related Microbiological Phenomena*, Academic Press, New York, 1978.
46. Ref. No. 7, p. 419.
47. T.N. Lung, The History of Copper Cementation on Iron in Medieval China, *Hydrometallurgy*, 17, 1986, pp, 113-29.
48. Ref. No. 7, pp. 431-2.
49. *Ibid.*, pp. 432-3.
50. *Ibid.*, pp. 256, 419-20.
51. A.K. Biswas, The Primacy of India in Ancient Brass and Zinc Metallurgy, published in *Indian Journal of History of Science*, 28, (4), 1993, pp. 309-30. Also *vide* chapter V in this book.
52. Lynn Willies, Ancient Zinc-Lead-Silver Mining in Rajasthan, India — Interim Report, *Bulletin of the Peak District Mines Historical Society Ltd.*, Vol. 10, No. 2, Winter 1987, pp. 81-123.
53. Lynn, Willies Ancient Lead-Zinc-Silver Mines in Rajasthan, *Mining Magazine*, January 1989, pp. 31-5.
54. I.C. Freestone, P.T. Craddock, K.T.M. Hegde, M.J. Hughes and H.V. Paliwal, Zinc Production at Zawar, Rajasthan, in *Furnaces and Smelting Technology in Antiquity*, ed. by P.T. Craddock and M.J. Hughes, Occasional Paper No. 48, British Museum, 1985, pp. 229-44.
55. A.K. Biswas, and Rakesh Kumar, Zinc Recovery from Zawar Ancient Siliceous Slag, *Hydrometallurgy*, Vol. 15, 1986, pp. 267-80.
56. B.B. Rao, and K.S. Rao, Lead Deposits in Varikunta-Zangamrajupalle Belt, Cuddapah District Andhra Pradesh, *Geological Survey of India Miscellaneous Pulication*, No. 27, 1977, p. 91.

57. Ref. No. 7, pp. 281-311.

58. Captain C.G. Dixon, Some Account of the Lead Mines of Ajmer, *Gleanings in Science*, Vol. III, April 1831, pp. 111-15.

59. Ref. No. 7, pp. 299-300.

60. *Ibid.*, pp. 182-3.

61. *Ibid.*, pp. 173-231. Also *vide Indian Minerals*, Volume 25, 1971, pp. 82-92 which provides 170 references on gold ore processing in India.

62. Ref. No. 3, p. 491.

63. Ref. No. 5, pp. 38-9.

64. Ref. No. 7, pp. 213-16.

65. S.R. Sarma, Varnamalika System of Determining the Fineness of Gold in Ancient and Medieval India in *Aruna Bharati: A.N. Jani Felicitation Volume*, Baroda, 1983, pp. 369-89.

66. R.E.M. Wheeler, and S. Piggott, Iran and India in Pre-Islamic Times, *Ancient India*, No. 4, 1947-48, pp. 85-103.

67. V. Ball, On the Building and Ornamental Stones of India, *Records of the Geological Survey of India*, Vol. VII, 1874, pp. 98-123.

68. Ref. No. 7, pp. 455-72.

69. Ref. No. 7, pp. 532-55.

70. B.C. Raymahashay, and S. Sharma, Decay of Building Stones: A Mineralogical Model for Konark Sun Temple, India, *Quarterly Journal of Engineering Geology*, 26, 1993, pp. 155-7.

71. Shihāb al-Dīn al-Umarī (d. 1348) English translation, I.H. Siddiqi and Q.M. Ahmad, *A Fourteenth-century Arab Account of India*, Aligarh, 1971; *Fakhr*-i-Mudabbir's work ed. by Ahmad Suhaili Khvānsarī, Tehran, A.H.S., 1346; S.D. Goitein, *Studies in Islamic History and Institution*, Leiden, 1966.

    Niccolao Manucci (1699-1709), *Storia do Mogor*, tr. by W. Irvine 4 Vols., London 1907-8. All referred in Reference No. 3, p. 77 and 333.

72. D. Havart, *Rise and Decline of Coromandel*, Original Dutch in 1692 or 1693, quoted by Dharampal (Reference No. 28, p. 57).

73. Captain Alexander Hamilton, Account of a Voyage to the East Indies, 1708, Vol. I, p. 395 quoted by S.G.T. Heatly, *Journal of the Asiatic Society of Bengal*, Vol. 12, 1843, pp. 542-3.

74. George Pearson, Experiments and Observations to Investigate a kind of Steel, *Philosophical Transactions*, Vol. 85, 1795, pp. 322-46.

75. Dr. Benjamin Heyne, Iron Works at Ramanakapettah, report sent to Governor of Madras, September 1, 1795, reprinted in Reference No. 28, pp. 263-7. Also included in Heyne's *Tracts, Historical and Statistical on India, with Journey through Various Parts of the Peninsula*, London, 1814.

76. D. Mushet, Experiments on Wootz, *Proc. Royal Soc.*, I, 1805, pp. 183-5.

77. J. Stodart, A Brief Account of Wootz, *Asiatic Journal*, V, 1818, p. 570. Stodart assisted Michael Faraday in analysing Indian wootz.

78. F. Buchanan Hamilton, *A Journey from Madras through the Countries of Mysore, Canara and Malabar*, London, 1807.

79. H.W. Voysey, Description of Native Manufacture of Steel in South India, *Journal Asiatic Society of Bengal*, I, 1832, pp. 245-7; II, p. 402.

80. W. Cracroft, Smelting of Iron in the Kasia Hills, *Ibid.*, I, 1832, pp. 150-1.

81. Major James Franklin, *Observations on Several Iron Mines in Central Part of India*, 19th May, 1835, India Office Library, Reprinted in Reference No. 28, pp. 269-94.

82. H. Yule, Notes on the Iron of Kasia Hills, *JASB*, XI, 1842, pp. 853-7; *vide* also Oldham, *Mem. GSI*, Vol. I, p. 201.

83. J.M. Heath, On Indian Iron and Steel, *Journal Royal Asiatic Society*, Vol. V, 1839, pp. 390-7.

84. S.G.T. Heatly, Contribution towards History of Development of Mineral Resources of India, Memoranda Relative to Working of Iron in Bengal, *Journal Asiatic Society of Bengal*, Vol. XII, 1843, pp. 542-6.

85. Capt. L. Jacob, Report on the Iron of Kathywar, *Journal Royal Asiatic Society*, Vol. VII, 1843, pp. 101-2.

86. V. Ball, 1881, Reference No. 7, pp. 335-414.

87. Dharma Kumar and Meghnad Desai, *The Cambridge Economic History of India,* Vol. 2, Cambridge University Press, 1982.

4

Iron and Steel in Pre-Modern India

A Critical Review

Introduction

THERE exist several critical reviews on iron and steel in ancient India.[1-7] The dates of the earliest iron age sites in India show that the discovery and use of iron started around 1200 BC. indigenously and independently in at least three nuclear zones — Karnataka in the south, U.P.-Rajasthan in the north, and West Bengal-Bihar area in the north-east.[6] In terms of surface carburisation, forging and lamination techniques, the Indian tradition has been as ancient as any other outside India.[2] The seventh-sixth century BC smelting traditions at Ujjain, Dhatwa and Naikund have been remarkable.[3,5] At Naikund, carburisation of iron used to be done to the extent of 1 per cent carbon in the ingot.[4,5]

The earliest evidence (obtained so far) of hardening through quenching and tempered martensitic structure, pertains to a third-century BC iron sickle from Pāṇḍu Rājār Dhibī. Ghosh and Chattopadhyaya have dated a sample of hardened steel at Barudih in Bihar to be 810 BC (*MASCA Journal* 2, 2,63, 1982). The issue raised by Bronson[7] as to whether India produced crucible steel before sixteenth century AD has been contested by Prakash.[6b] The subject is wide open for further research.

How exactly the corrosion-resistant forge-welded structures like the Delhi Iron Pillar, Dhār Pillar or the Koṇārk beams (Vide Plates 32, 33 & 36) were made, is not well-understood. The roots of the tribal traditions of iron-making in India, reported by several scholars,[6] must have originated in the ancient period before AD 1200; but the exact dates are problematical to find. Before we critically examine the late eighteenth/early nineteenth-century scenario, a few historical facts related to the medieval India may be noted.

Iron and steel in medieval India

The historical records pertaining to iron and steel in medieval India are scanty. Ray Chaudhuri and Irfan Habib have provided some information in this regard.[8] The mining areas lay scattered in the hilly region beginning with Gwalior and extending to the tip of southern India. The Geniza records of the eleventh and twelfth centuries show that the Deccan exported iron and steel to the Middle East.[9] Fakhr-ī-Mudabbir thought the Indian swords to be the best, and writes that the damascened sword (*maujdaryā*) was the rarest and fetched the highest price. Another kind was made of soft iron alloyed with copper and silver, and still another, from Kurij in Cutch was made of steel.[10] *Rasa-ratna-samuccaya* of the thirteenth century AD described different varieties of iron and steel, which shall be presented later.

Manucci reported that iron ore from Gwalior mines was smelted and the metal was finally processed in different Moghul provinces such as Bengal, Allāhābād, Āgrā, Berār, Gujarāt, Delhi, Indalvai, Indur and Kashmir. India exported over 10000 pounds of steel in 1657 and over 5 lakh pounds of ironware in 1667. The Dutch establishment shipped in the seventeenth century from India 20000 ingots of 'wootz' steel.

During the 1660s the Dutch were active in developing the iron industry in the Godāvarī delta. They were in such an urgent need of pig iron, iron bands, iron bars and cannon balls, that they organised a manufactory system of production for these items in their factories. Varied iron products such as nails, cannon balls, iron bars, iron bands, etc., used to be exported from Coromandel to Batavia. Spikes, bolts and anchors were produced in India for shipyards at Narasapur, Masūlipaṭam, Pulicat, etc. In the Qutbshāhī kingdom, Indalvai (near the modern Nizāmābād) was the centre of the manufacture of swords, daggers and lances which were made from iron that was mined in the Kālāghāṭ hills.[11-13]

Mining was widespread in Golkoṇḍā, Karnāṭaka and Mysore. Whereas Mysore iron and steel were of average quality and used mainly for ploughshares, horseshoes, nails, etc., for local consumption, Golkoṇḍā steel was famous all over

India and even outside the sub-continent.[12,14] Another important area of iron ore mining and ferrous metallurgy in medieval India was Assam. The craftsmen supervising smelting were known as *ojhās*. Double-bellowed furnaces in Jaintiā and other parts of the Khāsī hills were more efficient than those in other parts of Assam. The Moghul India used to import iron hoes and bell-metal vessels from the Ahom state.[15]

Ever since Bābur invaded India in AD 1526, artillery and gun founding started in the sub-continent. Most of the earlier cannons were made of cast bronze and brass. Some of these, such as *Malik-i-Maidan* (Vide Plate 31) used in AD 1564 were excessively heavy and difficult to transport. The Bījāpur kings requisitioned the services of many technicians from Turkey.[16] Gradually attention shifted towards the fabrication of iron cannons.

Amir Fathaullah, a reputed mechanical inventor, had come from Shirāz in Persia to assist Ādil Shāh of Bījāpur. During AD 1584-90 he served Akbar's foundry establishment with great distinction. New types of portable iron guns and matchlocks, *bandūkas* were made. A mechanically driven wheel was invented by the motion of which sixteen gun barrels could be cleaned in a very short time.[17]

Despite the presence of Turkish and European gun-casters in India, cast iron cannons were never introduced. During Akbar's regime the gun-barrel was made up of discs of wrought iron through which holes had been punched using hammers and chisels. These perforated discs were joined together by forge-welding and the joints were reinforced by slipping red-hot iron rings into these, so that on cooling, these rings contracted tightly on to the joints. (Vide Plate 37) 'The expertise in forging techniques was therefore diluting the need for casting technology'.[17,18] It is strange that despite the knowledge of refractory clay and the presence of foreign technicians, there was so much sluggishness in medieval India to pass on from bronze casting to the casting technology of molten iron (higher melting point of 1535°C).

The eighteenth-nineteenth century scenario: diversities

Now we proceed to record and critically review the writings of the European travellers on the iron and steel scenario in the eighteenth-nineteenth-century

India. This information can partially, though not fully, compensate for the miserable lacunae in the medieval literature on the subject. One is immediately impressed with diversities in ores processed, smelting techniques, furnaces, and products obtained in different geographical sectors of pre-modern India.

Valentine Ball recorded the diversity of iron ores mined, purified and processed by the Indian iron smelters,[19] mentioning the reports of the earlier workers.[20-26] We have briefly reviewed the literature regarding the mineralogical characteristics of different ore-bodies indigenously processed. Table 4.1 contains a summary.

In India three types of furnaces were used: bowl-shaped open hearths, small blast furnaces and tall blast furnaces.[28-31] The Choṭā Nāgpur small blast furnaces were typically 4' 6" high, the blast being introduced by a single twyer. The Mālābār district Stückofen type tall blast furnaces were typically 10 ft high. Several furnaces were built together over a common platform. Charcoal was the only fuel used and flux was rarely used. India produced three kinds of product: crude wrought iron, pure wrought iron and wootz steel. Kāṭhiāwār furnace was exceptional in so far as it resembled the European reverberatory furnace and was probably a technological import/adaptation. Flame produced by the blast over the burning charcoal played over the ore which was kept in a separate compartment, instead of being mixed with fuel. Thus reduction of iron oxide was by carbon monoxide rather than carbon.

In places of large scale production such as Mysore or Birbhūm, the refining forges were separate and the refiners bought the iron blooms from the smelters. Steel was made either by carburising wrought iron in crucible (*wootz*) or by decarburising cast iron.

Recently some studies have been made on the pre-modern technology by reconstructing ancient furnaces with the aid of the aged rural artisans, and measuring operating conditions and analysing raw materials and products.[6b, 32-35] These replication experiments and data have helped in quantification of the pre-modern technologies. Now we proceed to describe specific regional patterns in iron/steel technology specifically in the southern, central, eastern and western India.

Table 4.1

Indigenous processing of iron ores in pre-modern India

Location	*Type of Ore*	*Comments*	*References*
Assam (Upper)	Clay ironstone	3000 smelters in sixteenth-seventeenth centuries	Hannay in *JASB*, *25*, 330
Khāsī and Jaintiā Hills	Titaniferous magnetic oxide	Beneficiation by raking	Yule.[24,26]
Birbhūm in West Bengal	Partly earthy, partly magnetic	Early production of *molten* iron — early eighteenth century	Hamilton[20]
Bālāsore, Orissa	Lateritic ore	Some collaboration with the Dutch in seventeenth century	Hamilton[20]
Pālamau, Bihar	Three types of ore, Agāriās preferring the magnetic type	Details on smelting by the Agāriās.	*Mem. GSI*, Vol. XV, part 1, p. 112 Ref. No. 19, pp. 376-81
Sambalpur	Magnetite in metamorphic rock	Use of *śāl* wood charcoal	*Rec. GSI*, Vol. X, p. 182
Tendukhera Madhya Pradesh	High grade, slightly calcareous	Production of *kaccā* iron and *pakkā* steel	Franklin[25] Ref. 19, p. 386 Ref. 27, pp. 269-94
Gwalior	Rich deposits, now deserted	Very ancient tradition	Hacket, *Records GSI*, Vol. III p. 42. Ref 19, p. 394
Nimār and Mālwā in Indore Territory	Barwai deposit is chiefly brown hematite	Mentioned in *Āin-ī-Akbarī*	Ref. 19, p. 397

Location	*Type of Ore*	*Comments*	*References*
Kāṭhiāwār and Cutch	Laterite of sub-nummulitic group	Reverberatory furnace. Production of wootz (1838-40)	Ref. 19, pp. 400-2
Hyderābād	Two kinds of ores — lateritic and magnetic sand	Steel mines of Nirmal mentioned in *Āin-ī-Akbarī*	Voysey[23]
Cuḍḍapah and Kurnool	Siliceous hematite and also lateritic	Series of iron-smelting villages	1795 report of Heyne[21]
Mysore	Black sand iron ore in Venkatagiri	Manufacture of steel	Buchanan[22]
Cītaldurg in Sālem	Non-magnetic ochrous limonite	Manufacture of wootz	Heyne[21]
Mālābār or Kerala	Black magnetic sand	Rock 'laterite'-term first coined	Buchanan[22]

Pre-modern iron technology in southern India

During his famous journey through Mysore, Karnataka and Malabar (Kerala), Buchanan noted[22] several places where lateritic (the term coined by Buchanan himself: lateritis = brick) iron ores were processed and smelted. At Colangoḍu in Mālābār, four furnaces smelted black magnetic sand. At Velater there were 34 furnaces which belonged to a Mopla.

The lateritic ore was beneficiated by washing in a trough, open at both ends, which was placed in a running stream. The washed ore was largely magnetic. Buchanan provided admirably drawn sketches: sections and elevations of twin-shaft (250 kg iron per day) furnance which according to Prakash 'conform to the design parameters of the modern 1000 TPD furnances'.[6b,29] We do not know whether the local artisans had been assisted during the earlier centuries by some Portuguese or Dutch travellers. These 10 ft tall furnances were provided with bellow-pairs, peep holes, hearth and slag pit. Dirar and others have recently attempted to revive the ancient process in Kerala.[33,34]

A report on the manufacture of bar iron in the Salem region of south India was recorded by Campbell[36] in 1842 reproduced by Rahman, Subbarayappa[37] and Dharampal.[38] The furnace was constructed with common red potters clay. On a square platform of about 2 ft dimension with 5 inch thickness the furnace wall was 4"-6" thick. The annular space inside was partly cylindrical (9" diam 18" ht) and above conical (18" ht and diam tapering 9" to 3") and then expanding like a funnel up to 9" diam. The blast pipe was a cylinder of dry clay 14" long, about 4" thick, pierced with a hole of one inch in diameter. It was introduced into the furnance 5" above the bottom. The blast pipe was connected to the bellows made of two whole goat skins.

The furnace was filled to the neck with charcoal, about 26 pounds, and fire introduced. Then the common magnetic iron ore of Salem (about 5 lb) was mixed with charcoal (10 lb) and introduced into the furnace.

Normally an 11 lb lump was obtained as the product, one-fourth of it being recovered in the forge as iron. Campbell wrote:

> As regards their quality the worst of them is as good as the best English iron and its supposed defects arise from its almost always containing a considerable portion of steel. . . . A fractured end of a bar of iron presents a very different appearance — no glistening portion like English bar iron: it is granular or an aggregation of crystalline grains.

Campbell specifically mentioned that the south Indian iron-making furnaces resembled the 10-15 ft high, 3 ft diam ancient Stückofen furnaces in Germany. Thus, some Dutch influence in the design of this kind of furnace cannot be ruled out.

Buchanan made frequent allusions to the manufacture of iron in different parts of Mysore, and gave detailed accounts of the process. Many of his observations[22] as well as those of Heyne[21] were made in connection with the manufacture of steel, and hence shall be described later. In the Shimoga district, iron ores were worked in some parts. Magnetic ore occurred at Kodachadri in which is located the famous 10 m iron pillar. An effort was made during the middle of the nineteenth-century to tabulate the number of mines and furnaces for iron working in the districts of Shimoga, Kadur and Cītaldurg. Ball commented that 'it is incredible that in the years between 1872 and 1875 there were more than

1400 mines'.[39]

In the Andhra Pradesh area, Cuḍḍapah and Kurnool districts had many iron works. A series of iron-smelting villages lay along the eastern side of the Khundair valley (8 miles north of Dhoor). At a small village called Colapeṭṭa there were ten furnaces. In other parts of the Cuḍḍapah district there were furnaces, and itinerant blacksmiths carried their implements and wandered over the district seeking for employment.

Manufacture of iron in the Kṛṣṇā and Godāvarī districts attracted notice for many years. In September 1795, Banjamin Heyne referred to his visits to Lutchmiporum (Laxmipuram) and Ramanakapeṭṭāh iron works, the latter being very close to Noozeed and the diamond mines of Mallavilly.[40] Before the terrible famine of 1790-92, Ramanakapeṭṭāh was in a state of affluence with a great number of silver and coppersmiths and 40 iron-smelting furnaces; after the famine the number came down to ten. Each furnace required nine men, chiefly employed in working the bellows, 'an operation which might be easily improved by a proper mechanical apparatus, through the medium of fire or water, or both'. Preparation of steel at Nirmal and Konasamudram in the Nizām's territory would be discussed later.

Iron in the pre-modern central India

Several ancient iron ore mines were known to exist in the districts of Jabalpur, Baragāon, Pannā, Katola and Sāgar. Franklin described in around 1827 these mines and the mode of manufacturing iron in central India.[41] The Tendukaira or Tendukhera mine and iron works in the Sāgar District were described in great detail.

The smelting furnaces were 'very exact in their interior proportions and constructed with precision'. Franklin observed that the production of four furnaces (per week) was 71 *mauṇḍs* or 2.25 English tons, cost of production being two pounds six shillings per English ton.

Part of the bloom used to be forge-hammered just short of fusion, then 'rolled in the ashes of burnt cowdung, previous to fresh hammering, heated red-hot and plunged into water'. It was then an excellent steel. Franklin made suggestions with reference to the introduction of simple machinery in the works. In the year

1830, Colonel Presgrave opened a suspension bridge over the Bias river in Sāgar, the iron for which had been all smelted at Tendukhera.

In 1857 there were about 70 to 80 furnaces in Tendukhera. The charcoal was made in the jungle and *kaccā* iron blooms made in Catalan type furnaces. The *pakkā* iron was made in a different form, from which it came in the condition of crude steel, which was hammered for making edge-tools.

Ball recorded that the 250 ft high hill Khaṇḍeśvar (near Dewalgāon) in the Chanda district was wholly laden with iron ore and in view of the abundance of fuel available, 'this hill alone might furnish the whole of India with iron'. [42] P.N. Bose the first Indian geologist of the modern era recorded good quality iron ores in the Chanda district. Bose saw ancient mines in the western portion of the district at Raipur: mine of Magarkuṇḍ, furnaces at Koṇḍkasar, Dalli, Nandgāon, Khairagar, etc. Flux was never used in the furnaces and wooded forests near Dalli were so plenty that large charcoal furnaces seemed to be quite viable for long-term operation. The Agāriās who were working the furnaces were 'a very unsettled people, leaving a place as soon as the neighbouring jungle failed to satisfy their requirements, or the *zamīndār* enhanced the duty levied on thier furnaces'.[43]

Elwin surveyed the whole of central India and made ethno-technological studies on the iron-making by the Agāriās.[44] Krishnan has also reviewed the status of iron-making in ancient India.[45] The Muṇḍiā and Halbi tribes of Bastar were well-versed with the iron-making technology and they were capable of producing forgeable iron from their tiny furnace. The region of Bastar could not earlier be surveyed and included in the 'Agāriā' belt 'because of the inaccessibility of that region or due to the traditional self-containment of these tribes'.

Prakash and Igaki surveyed actual performance of the ancient iron making in the Bastar district.[46] The furnace used by the tribes was essentially of the same type as that used in the Chiglabecha area of Korapūṭ.[47] The bowl-shaped furnace was built below the ground level by digging a square pit of about 2000 × 2000 mm and 500 of mm depth and with steps on one side to go down. The furnace had a total height of about 800 mm and the diameter of the furnace at the throat 200 mm. The shaft of the furnace was 600 mm in depth and had a tapered wall. It ended into a bowl-shaped hearth having about 240 mm maximum diameter and

100 mm depth. A hole was provided in the bowl to tap out the slag. The furnace was made below the ground level to cut down the access of air and inner wall was coated with a layer of refractory clay.

The air blowing rate wad adjusted to maintain a temperature of around 1150°C. The whole operation took 5-6 hours producing one kg of iron. The semi-fused mass of the sponge iron block mixed with slag was taken out and separated from the solidified bottom charge. The sponge iron block was forged on an anvil to flatten it and squeeze out the slag, the response to forging giving an idea of its carbon content.[46] The Muṇḍiā and Halbi tribes have been living a life of self-sufficiency and making iron by the ancient method to prepare arrowheads, axes and hoe which form a part of their routine life.

Recently Balasubramanian, Ballal *et al.* have documented[33] some typical ancient iron-smelting furnaces in Madhya Pradesh — used by the Lohra family at Ghaṭgāon (Dist. Rāigarh) and Wadruffnagar (Dist. Sargujā). The smelting furnace in the unbroken tradition of Ghaṭgāon is a cylindrical clay structure 600 mm in diameter and 950 mm in height on the outside. The shaft inside extends 150 mm below ground level so that the total shaft height is 1100 mm. The shaft can be divided into 3 sections: (a) lowermost cylindrical portion of 250 mm diameter and 550 mm height, (b) the next section a truncated cone with a base diameter 250 mm, top diam 100 mm height 250 mm and (c) top cylindrical section of 100 mm diam 300 mm height.

The furnace used by the Agāriās from Wadruffnagar is a squat structure, 600 mm above the ground level. The shaft extends to 270 mm below the ground. The diameter is 120 mm at the top. The shaft runs straight down at the front but opens out at the side and rear to a diameter of 300 mm over a length of 450 mm. The lower portion is a cylinder of diameter 300 mm.

Locally available clay is used for the construction of the furnace. Vegetable matter like Kodon chaff, rice husk or hay may be mixed with the clay for reinforcement and stabilisation. A sloping pan is used for beneficiation of ferruginous sand from the river bed by gravity separation. The lighter siliceous particles are washed away with water and the heavier black ore particles, left behind, contain higher per cent of iron. These are sun-dried and put into the furnace with local charcoal.

Pre-modern iron making in eastern India

There were certain specific features in the iron-making traditions prevalent in the eastern India. The smelting of iron was an important industry in upper Assam, specially in the Nāgā Hills. According to Colonel Hannay (*Journal of Asiatic Society of Bengal*, Vol. XXV, p. 330), there had been at one time 3000 smelters and smiths in upper Assam. The principal centres of their operations were at Tirugāon and Hattigar, where the remains of their slag heaps and furnaces are still to be seen. During the sixteenth and seventeenth centuries, Assam was famous for the manufacture of big iron guns; one at Raṅgpur fort measured 17 feet 3½ inches and the metal was 7¾ inches thick. The casting technology could have been borrowed from the Chinese. This tradition died down on account of the Burmese incursions and competition from the Khāsī Hill technology. Hydraulic mining and beneficiation of the iron ore sands in the Khāsī and Jaintiā Hills have been described earlier.[48,49]

F.R. Mallet observed[50] that the smiths of the upper Assam used contrivances quite different from those used in India proper. In most parts of the sub-continent bellows were used: the supply of air to the furnace was provided by alternate expansion and contraction of a flexible material, like leather or raw skin. The contrivances used in upper Assam were made of blowing cylinders — single-acting or double-acting made of rigid material such as solid piece of wood, turned or bored. Into the ends, discs of wood were fitted by rough dovetailing, the circular joints between discs and cylinder being rendered air-tight with clay. In the centre of one disc is a small hole in which the piston rod works; the latter is made from a slip of bamboo split from a large-sized piece. Close to each end of the cylinder there is a hole of an inch or so in diameter, which allows the air to pass into the tubes or the tuyere. Thus the system worked like a blowing machine which substantially raised the temperature of the charcoal-fired furnace.

The double-cylinder machines were observed at Burhāṭ on the Disāṅg. Modifications were noted in the Naga Village of Rangkatu, 11 miles south-east of Mākum. Quoting Percy,[28] Mallet noted that the wooden blowing machines were common amongst the Chinese and the Burmese, and it is probable that the Assamese had acquired a knowledge of the principle from them.[50]

The existence of iron in the districts of Bālāsore in Orissa and Birbhūm in

West Bengal seems to have been known at the earliest period of British rule in the country. Captain Alexander Hamilton narrated in 1708 how he found on his way from Bhadrak to Bālāsore 'iron so plentiful that they cast anchors for ships in moulds'.[51] Is this occurrence of cast iron attributable to the earlier Dutch influence?[52]

In Birbhūm, Hamilton found 'iron ores carried by the *beparries* to established markets called *aurańgs*, where it was purchased by the smelters, whose furnaces or *sāls* adjoined the *aurańg*'. The Rājāh's officers supervised the operations and products of the *loha-mahal* having *kot-sāl* or roasting furnace and *khamar-sāl* where the iron was finally prepared for use.[51-53]

Sanyal reviewed[54] two types of indigenous pre-modern iron industries operating in Birbhūm during 1843-81. The first type (I) had small furnaces operated by the itinerant tribal people like Santals/Muṇḍa Kols.

The second type (II) of manufacture was carried out by the Bengālī-speaking people in the north-western part of the district, where ore as well as timber for charcoal fuel were easily available. Large scale furnaces being operated here were not to be found anywhere else in India. Seventy furnaces in the centres of Balliā, Nārāinpur, Deochā, Dumrā, Ganpur (Chowdhuries were the proprietors), etc., produced in 1852 (when Dr. Oldham visited the area) more than 2380 tons of *kaccā* iron. At Deocha itself there were 30 furnaces; average annual output from one furnace was 34 tons. In these furnaces the *kaccā* iron, 'unlike that produced in other parts of India, formed at the bottom of the furnace in a *molten condition*, and resembled good pig iron' (*Engineers' Journal*, Calcutta, Vol. III, 1852, p. 112). The refining was really a sort of puddling process, which induced a pasty condition admitting of the iron being drawn out and hammered until it became thoroughly malleable. The Muslims supervised the production of the *kaccā* iron and the Hindus refined them to *pakkā* iron.

Dr. Oldham observed that this soft charcoal iron had its uses on account of malleability, but in price it could not compete with English iron at the prices then prevailing in Calcutta. Absence of economical fuel (forest wood charcoal) made the future of the process quite bleak.[55] During the previous century, Indranarain Sarma, an Indian, wanted to produce iron[53] at a large scale in Birbhūm (1774). During 1777-89, Farquhar had used the indigenous process — not the European method — to cast shot and shells and hook iron and supply them to Fort William[55] prior to switching over to the gunpowder manufactory at Phulṭā (1789).

Hamilton's 1708 note on iron in Orissa has been already mentioned.[51,52] Not much detail is available regarding the late seventeenth-century foreign influence in the local technology. In 1964, Moni Ghosh submitted his report on the primitive iron-making in the Korapūṭ district of Orissa.[47] These furnaces are about 30" high, height of the semi-spherical dome from the bottom of the furnace is 17 inches, height of the chimney 13", and the internal diameter of the chimney 5.5". Ghosh noted two types of furnaces in the area. The first is a shaft furnace and is fully subterranean. The second, which is partly subterranean is constructed over a rectangular pit.

Ball recorded the richness of the iron ore deposits in the Pālamau sub-division of the Lohārḍāgā district — three distinct types (a) magnetite, (b) siderite and hematite, and (c) red and brown hematite — and the large variations in the smelting techniques adopted by the Agāriās. One typical account was described.[56]

The furnaces of the Agāriās were built of mud and were about 3 to 4 feet high, tapering from a diameter of about 2½ ft at base to 1½ ft at the top. The hearth was a rounded cavity, which was about 10 inches in diameter, and the circular shaft above it is 6 inches in diameter. A bed of charcoal having been rammed down the hearth, ignited charcoal was placed above it, and the shaft was filled with charcoal. The blast was produced by a pair of kettledrum — like bellows, consisting of hollowed drums of wood, with goats' skins attached, and nozzles of bamboo. The bamboos which convey the blast were luted into clay tuyeres, which were themselves luted into the front of the furnace. The blast used to be kept up for six hours. (Vide Plate 10)

Powdered ore was sprinkled in alternate layers with charcoal — on the top of the ignited charcoal in the shaft. The slag formed was tapped through a hole. During the end of the process, the clay-luting of the hearth was broken down, and a *giri*, or ball of semi-molten iron, including slag and half-burnt charcoal was taken out. This was hammered to squeeze out the molten slag. The iron was then either refined by the Agāriās or sold to the *lohār*s who worked it up into bars.[56] If the hearth were larger, bloom produced would also be large and unworkable. Such a big piece (156.5. lbs) was worshipped for many years by the people of the

Kharakpur Hills in the Monghyr district![57]

Moving west from Bihar, we find ancient iron-making traditions existed in Gurgāon District, Haryana. In 1831, the Ferozpur/Narnul ores were being smelted in the smelting furnace called Māndarī which was merely a square receptacle leaving an opening both above and below; the dross ran out from the lower orifice into a hollow sunk into the ground. Two bellows were used for each furnace. Iron obtained as an imperfectly melted conglomerate was forged for heavy wedge-headed hammers.[58]

Iron in pre-modern Gujarat area

In the Year 1840, Captain Grant described the manufacture of iron in Cutch which was conducted according to the ordinary Indian method. It was reported that *wootz* or steel used to be produced there formerly.

The iron industry in Kāṭhiāwār was described by Captain G. Legrand Jacob in 1838 (*Selection from Records*, Bombay Government, No. XXXVII, p. 465) and this description was quoted by Ball.[59] There were six groups of furnaces, of which two, situated respectively at Ranawao and Ranpur, were visited by Captain Jacob.

The lateritic ore was smelted in the furnaces of unusual shape; these were 'something of the nature of a reverberatory furnace, except in the essential particular that the ore was placed in partial contact, though not mixed with the fuel'. The furnaces were long and narrow (dimensions not given), and were built of brickwork lined with clay. At one end there was a chimney, and close to the other extremity there were two apertures on opposite sides; through one of these the blast from the bellows was driven over the ignited fuel, and by the other the slag was withdrawn. The ore was placed in hollows on a bed of charcoal, and was covered with a layer of the same, but was so arranged to be on either side of the stream of air.

The bellows consisted of two bullock hides sewn round bamboo hoops in vertical rings and worked alternately by downward pressure, the operator closing the mouth of the sack with his hands as he weighed upon it with his chest and arms. The furnace held little more than seven Bombay *maunḍs* (1 *maunḍ* = 40 *seers*), which took from six to eight hours to melt.

Captain Jacob calculated that the metal recovery was about 40 per cent. The workmen said that the direction of the wind affected the outturn, 'five *mauṇḍs* being obtained from a double charge of the furnace, during westerly winds, and seven *mauṇḍs* when it blew from the east'. 'This they accounted for by saying that metals were like mortals, some climates agreeing with them better then others'. Jacob and Ball attributed the difference to the comparative vigour of the men who worked the bellows under the influence of moist and dry winds. It is equally possible that moist wind resulted in lowering the furnace temperature.

Captain Jacob wrote that 'it was difficult to witness without pain the struggles of these men for a subsistence, which, owing to competition, was yearly becoming more arduous'. He predicted rapid extinction of this industry which actually took place in a few decades since 1838. Ball reported in 1880 the *new modern* blast furnaces in India started by the Governor General for the Rājā of Sirmur at Nāhan.

Replication experiments and theoretical studies on pre-modern technologies

Moni Ghosh reported operations on the ancient Korapūṭ furnaces in Jiragara (Bihar) and Chinglebecha and also in Kamarjoda near Jamshedpur.[47] The iron smelting practice at Bishunpur has been simulated by Vikas Bharati, a voluntary organisation with the help of Nagu Birjia, an old technician of this area. Prakash has compared the operating parameters of Jiragora and Bishunpur (Dist. Gumla, Bihar) iron-making furnaces and also compiled composition of iron produced by ancient Indian furnaces.[6b] The compiled data are reproduced in Tables 4.2 and 4.3. Balasubramanian, Ballal *et al.* have described[33] the operating features of the ancient furnaces at Bishunpur (Bihar) as well as at Ghaṭgāon (Dist. Rāigarh) and Wadruffnagar (Dist. Sargujā) in Madhya Pradesh.

We have noted that most of the indigenous furnaces have varied height between 0.6 to 1.2 m and diameter 0.25-0.5 m. Charcoal from hard woods like *śal*, teak, etc., were used. Normally no flux was used and the blast air was at room temperature. Vivek tried to simulate this process at IIT, Bombay and we quote some of his experimental results and theoretical calculations.[35]

Vivek constructed the clay furnace 1.2 m high and the diameter varying from

Table 4.2

Comparison of operating parameters of Jiragora and Bishunpur iron-making furnaces (taken from Prakash, Ref. No. 6b)

Operation	*Time*	*Material Charged*		*Blowing Rate*	*Remark*
		Ore	Coal		
JIRAGORA FURNACE					
1. Furnace preheating	1h	—	8 kg	Faster than 87 strokes/min	
2. Reduction stage	4h	—	14 kg	87 strokes/min	Sequence of ore and coal char-ging
3. Consolidation stage	1h	—	8 kg	Vigorous blowing	not available
4. Total operating time	6h	—	—	—	—
5. Total material charged	—	24 kg	30 kg	—	5.6 kg of iron
BISHUNPUR FURNACE					
1. Furnace preheating	1.25h	2.5kg	3.25kg	40-50 stroke/min	Furnace Temp. 550°C
2. Reduction stage	4h	8kg	16kg	60-70 stroke/min	950°C
3. Consolidation stage	0.5h	1kg	2kg	110 stroke/min	1500°C
4. Total operating time	5.75h	—	—	—	—
5. Total material charged	—	11.25kg	21.5kg	—	2.5 kg of iron

Table 4.3

Chemical composition of iron produced by ancient Indian furnaces

Source	*C%*	*Si%*	*Mn%*	*P%*	*S%*	*Others*
1. Delhi Iron Pillar	0.23	0.66	—	0.18	Traces	N_2=0.0065%
2. Bhubaneswar Iron Beams	0.27 to 0.45	0.05 to 0.11	Traces to 0.04	0.015 to 0.018	0.006 to 0.015	Cr = 0.9% Ni = 1.6%

3. Bastar Iron Axe (100 years old)	0.25 to 0.45	—	—	—	—	Other elements in traces
4. Jabalpur Iron (recent)	0.59	110 ppm	40 ppm	—	—	Cu = 340 ppm Ni = 353 ppm Other in traces
5. Bishunpur Iron (recent)	0.016 to 0.043	—	0.057	0.02 to 0.2	0.007 to 0.013	—

0.3 m at the bottom to 0.15 m at the top. Air was supplied by simulated bellows, and charcoal from the local market was used. The experiments were conducted in summer months with 30°C temperature and 50-60 per cent relative humidity.

The Raceway Theoretical Flame Temperature (RAFT) was calculated to be 1500-1510°C if the input dry air temperature varied between 20 to 30°C. The effect of humidity on the RAFT was considerable. During summer, 30°C air with 50 per cent RH is supposed to give RAFT of 1400°C whereas during winter, 20°C air with 25 per cent RH would give a RAFT of 1475°C. (This also corroborates Captain Jacob's observation at Kāṭhiāwār[59]). Using a pyrometer 2 hr after starting the furnace, the maximum temperature was found to vary between 1400-1475°C. Through temperature measurements at different parts of the furnace, a thermal profile was constructed.

Significant reduction of the ore commenced only after it descended about half the height of the furnace. The particle of iron was molten for a very short time and that too on the surface. 'Bloom' is the aggregate of iron particles sintering into a highly porous mass with interstices filled with molten slag. In the modern blast furnace the liquid metal product picks up large proportions of C, Si, P, etc. In the ancient process, carbon content in the solid metal was low (α-ferrite has m.p. 1410°C) unless larger proportion of charcoal was used. Silicon, sulphur and manganese pick-up were also little. Charcoal being much purer fuel, impurity pick-up in the malleable metal was very low. Only the phosphorus part went into the metal, providing corrosion-resistance to it.

Since the lime-flux was rarely used, the slag was often fayalite, $2FeO, SiO_2$, melting in the range 1150-1200°C. The slag coating the metal further retarded carbon pick-up. In the ancient process considerable amount of iron oxide remained in the slag unreduced, iron recovery was low, but the purity of the metal was high.

Bloomgren and Tholander have shown[60] that high oxygen activity of the fayalite slag acted as a control parameter in lowering the carbon content of the metal. They have also shown that in the modern blast furnace, productivity and temperature being high and limestone being used as flux, fayalite concentration in the calcium silicate/calcium alumino-silicate goes down, oxygen potential decreases and hence the carbon pick-up in the metal is increased.[60]

Thus the indigenous technology had low productivity and thermal efficiency and yet produced pure malleable metal which was used to forge a large number of locally useful materials such as arrowhead, dagger, knife, spade, etc., and preferred over hard but brittle high carbon imported iron.

The charcoal crisis and the atrophy of the indigenous industries

Various reasons have been ascribed to the atrophy and the extinction of the indigenous iron industries in India. Neither the Moghul government nor the people in the country were aware of the new technological developments in the field of mass production of iron and steel, which were spreading fast in Europe.[31] The nineteenth-century European observers consistently remarked about the lack of innovation in the Indian industries, the persistence of charcoal furnaces, the non-use of mineral fuels, the non-use of flux, etc. On the other hand, some scholars believe that the most important factor has been the British competition, intrusion into the Indian market and ruthless elimination of the Indian industries. Dharampal has written:

> The disappearance seems to have resulted mainly from large-scale economic breakdown and hostile state policy.[27]

The phenomenon was indeed complex. One technical factor was the charcoal crisis which Bhattacharyya has deliberated upon in his masterly paper.[30]

Enormous quantities of charcoal was needed in the indigenous processes. The proportion of charcoal to that of iron yielded was often 5:1. Considering the refining stage, all told about 14.5 tons of charcoal was used to produce one ton of refined wrought iron. It has been estimated that to produce 2500 tons of refined wrought iron, 140,000 tons of wood were required which could be obtained from 437 square miles of heavily wooded land! Production of charcoal from green wood was so wasteful. Bhattacharyya rightly commented that

> the rapid depletion of fuel source seems to have been one of the causes of the atrophy of the iron-smelting industry in the nineteenth century.[30]

Europe had faced similar problem in the sixteenth century when forests were being depleted on account of the charcoal-based process in the iron industry. An acute shortage of timber threatened the ship-building industry in England. The legislative restrictions on waste of timber and high price of fuel charcoal led to technical experiments with coal as a substitute for charcoal. Abraham Darby introduced coke for smelting iron in 1709. The use of coke spread rapidly in England, but it is noteworthy that as late as 1806, 11 out of 173 pig iron furnaces were still using charcoal.

One reason why coal or coke could not replace charcoal as fuel readily, was that the conventional bellows were not sufficient to bring the mineral fuel to the required temperature. The coke furnace needed an engine two and a half times more powerful than that needed for a charcoal furnace for iron smelting. Charcoal could not be replaced with mineral fuel without auxiliary innovations such as superior blasting machines in place of the low-powered bellows.

At this stage, we may record a brief note on the use of coal and coke in ancient China. The Chinese people had been using efficient double-acting bellows since fourth century BC, the hydraulic bellows from the first century AD, and good refractory clay and surface coal from *c*. AD 400. Thus, efficient use of coke or coal demanded powerful and mechanised bellows which India did not have, except in the Chinese border of eastern India. China could develop furnaces having higher temperatures which melted iron, but they were aided by their possession of high-phosphorus iron ore which substantially reduced the melting point of iron and enabled production of cast iron.[61]

Thus, the Indian persistence of charcoal furnaces and the non-use of mineral fuels, the non-use of flux, non-availability of mechanised bellows, non-development of high-grade refractories represented different aspects of a complex phenomenon resulting in stagnation and atrophy of the indigenous craft. When the Western visitors attributed the atrophy to the Indian artisan's indolence, indifference and resistance against technological improvements, they were a bit too harsh and uncharitable. There were several examples when the British entrepreneurs tried coal or coke in indigenous furnaces and failed on account of the technical reasons cited earlier. (Vide Plates 6-8: early petroleum well and coal mines).

On the question of improving iron manufacture in India during the nineteenth century, there emerged two schools of thought. One view, supported by Bhattacharyya,[30] and now endorsed by the present author, was that the new technology had to be introduced — employing coke, good refractory, flux, mechanisation, etc., — *all at once* instead of making piece-meal improvements in the indigenous method of production. The other view propounded by M. Fryar, etc., was that the better method was 'to gradually engraft improvements on the native methods of manufacture'.[30] Dharampal felt that indigenous technology was hurt by 'unthinking transplanting of European sciences' and that the disappearance of the indigenous iron industries in India was on account of unequal British competition, 'large-scale economic breakdown and hostile state policy'.[27] Dharampal hardly attended to the technical considerations of the charcoal issue (as Bhattacharyya did). His observations had political overtones and would be discussed at the end of this article.

Iron-carbon alloys in India

The thirteenth-century text *Rasa-Ratna-Samuccaya* contains a section (5.69-5.96) on different varieties of iron.[62] The descriptions clearly indicate that Indians were familiar with iron-carbon alloys and different phases.

Kānta-loha is soft wrought iron with five sub-varieties of different magnetic properties.

The name *muṇḍa-loha* (second major variety) probably originated from *muṇḍiā*, one of the metalsmith tribes in the district of Bastar whose ancient iron-making process has been scientifically observed and recorded by Prakash and Igaki.[46] *Muṇḍa-loha* had three sub-varieties: *mṛdu*, soft which is *drutadrāva* 'melting or softening quickly' and could be low m.p. grey cast iron; *kuṇṭha*, expanding with great difficulty on hammering, could be mottled grey iron; and *kaḍāra* sub-variety which 'breaks on hammering' could be white cast iron. The indigenous metalsmiths produced the last two brittle varieties unintentionally, when temperature of the furnace rose high and amount of charcoal used was large, resulting in high-carbon iron. They preferred the first variety *mṛdu* which had smooth surface, did not contain fissures and was adjudged to be 'the best of the three'.

Tīkṣṇa-loha or carbon steel had six sub-varieties. These appreciably hard materials were carburised iron which could be hypo-eutectoid (less than 0.83 per cent carbon) or hyper-eutectoid (more than 0.83 per cent carbon) steel. *Pogara*s mentioned in the text (*RRS*, 5.77-83) meant hair-like lines which may be surmised as white cementite streaks, coarse enough to be seen by the naked eyes.

The *khara* sub-variety of *tīkṣṇa-loha* is free from *pogara*s (or has *pogara*s too fine to be seen), breaks on bending and has 'mercury-like shining fracture surface'. The *sāra* sub-variety looks fibrous. The other four varieties were probably the products of hot forging and quenching. *Hṛnnāla* looked whitish black and contained *pogara*s; this was *chedane atiparuṣam*, very hard to pierce. *Vājira* looked bluish black and was 'full of hard *pogara*s and overspreaded by dense thin linings'. *Tārāvaṭṭa* developed good cutting edge. *Kālāyasa* was described as 'bluish black in colour, dense, smooth, heavy and bright in appearance, whose sharpened edges do not get spoiled even by hammering'. This description is close enough to that of the damascened steel swords 'of a dull blue colour marked with ten millions of meandering lines' which are compressed and elongated cementite grains or streaks in pearlite-graphite-martensite matrix.

Wootz, a crucible steel of India

The Indian crucible process for making high quality steel might have started before the Christian era as evidenced by the excavations of crucibles dated 250 BC in Tamil Nadu.[46] Even earlier, quality steel of India was presented by King Porus to Alexander (323 BC) and Kauṭilya's *Arthaśāstra* referred (II.17.14; IV.1.35) to *vṛtta* or steel. The word *vṛtta* means circle or disc, and could denote crucible-molten steel, *ukku* in Telugu or Kannada, *wuz* in Gujarati or *wootz* which was the term used, and spread all over the world by traders from the Middle East.

In his excellent review,[7] Bennet Bronson grudgingly concedes that 'India may have invented or perfected from the second century AD the idea of crucible steel making, but that should not be accepted as a given in medieval studies'. Bronson felt that *wootz* was merely carburised iron showing brittleness; there were several processes for making it, some practised outside India; the art of converting *wootz* to steel objects such as Damascus sword was known and probably perfected outside India.

Prakash has contested[7] Bronson, and has recorded circumstantial evidences, such as the texts of *Bṛhat-saṁhitā* of Varāhamihira and *Rasa-Ratna-Samuccaya* (quoted by us earlier), etc., to prove that India was indeed the pioneer in producing different kinds of *wootz* and the end products of malleable steel. We also feel that *Rasa-Ratna-Samuccaya*'s 'bluish black' *kālāyasa* (5.82) which had been mentioned much earlier in Kauṭilya's *Arthaśāstra* (II.17.14), corresponds beautifully with the Damascus sword, 'the sharpened edges of which do not get spoiled by hammering', and thus establishes India's primacy not only in *wootz* but also in the finished steel. We have discussed our views on iron and steel in ancient India elsewhere.[1]

The earliest Western report on pre-modern India iron and steel was recorded by Havart.[63] Some of the earliest writings by the British travellers[20-26] have been already quoted. A few researchers specifically investigated *wootz* and Indian steel — both the processes and the products.[64-69] Stodart for example, sustained his interest on *wootz* right from 1795 to 1820, collaborating with Michael Faraday. These two collaborators tried to prove that "*wootz* was in reality an alloy of iron with aluminium or various other metals".[68]

It appears that there were *two* distinct processes for making *wootz*. The south Indian processes (there were minor variations), practised at Mysore and its surroundings, Tamil Nadu and Sri Lanka, corresponded to carburisation of wrought iron in crucibles. On the other hand, the Hyderābād process probably involved co-fusion in crucible of wrought iron and white cast iron.

South Indian processes

Heyne[21] and Buchanan[22] described several pre-modern south Indian processes for making steel. Heyne witnessed a process starting with ochreous limonite in a small village among the hills south-west of Cītaldurg. He also saw steel being made at Kakerahally, on the road from Bangalore to Seringapatam, using megnetic sand. Buchanan gave detailed accounts of steel-making processes in different parts of Mysore. He saw furnaces at Venkatagiri, Ghettipura (two *coss* from Magadih), Madhugiri, Chinnarayandurga, Hagalawadi and Devarayadurga. The Sri Lanka version of the process was described by Coomaraswamy.[70]

The crucible was usually made from locally available clay, mixed with carbonaceous materials like the rice husk (charred or uncharred) in equal proportions. CVB mentioned[71] the use of red clay, strained and purified by sedimentation. After moistening, the mixture was well mixed 'under the feet of oxen'[22] or by pounding in a stone.[71] The well-mixed clay was fashioned into crucibles of conical shape, size ranging between 250 ml to 500 ml. Coomaraswamy gave[70] the dimensions of the crucible as 200 mm long × 50 mm diam × 6.25 mm thick.

The crucible was filled with 250-500 g of bloomery iron, wood chips from *Cassia Auriculata* (*Awaram* in Tamil), leaves or creepers of plants, rice husk, etc. The crucible was then sealed with a sun-dried clay lid or a ball of moist clay. Coomaraswamy mentioned a hole in the cover for the gases to escape.

The closed crucibles were allowed to dry and then heated in a charcoal fire. The details of the furnace were given by Buchanan.[22] The furnace consisted of a horizontal ash-pit and a vertical fire-place, both sunk below the level of the ground. A row of crucibles was first laid round the sloping mouth of the furnace; within these, another row was placed, and the centre of the dome so formed, was occupied by a single crucible, making fifteen in all.

Air was blown with hand bellows usually made from buffalo skins; Commaraswamy's report shows however foot-operated drum bellows as used in central India. The number of crucibles fired at a time varied between 6 and 59.

The crucibles were then covered with charcoal and the fire started. The fire was kept up for about 5-6 hours, till the contents were molten, Charcoal was replenished periodically both above and below the crucibles. At the end of the operation, the crucibles were allowed to cool by throwing water over them or in the furnace itself. The contents solidified inside, taking the conical shape of the crucible. The top surface showed striations indicating crystallisation from the liquid state.

Rao, Mukherjee and Lahiri found[72] that the unused crucibles in the south

Indian sites such as Salem consisted mainly of aluminium silicates and free quartz. The fired crucible was found to be porous and contained mineral mullite, indicating that it had been heated to a temperature not less than 1300°C. Presence of cristobalite showed that the temperature might have occasionally risen over 1470°C.

In the ingot, the carbon content in the periphery was around 0.8 per cent and the value decreased towards the centre. This indicated diffusion of carbon from the organic content of the crucible and incomplete fusion. Bronson commented on the

> somewhat slapdash nature of the south Indian process by comparison with the rigorously controlled methods in use in Hyderābād.[7]

The Hyderābād or Deccani process

Wootz from Hyderābād was better suited for making Damascus blade and prepared by a different method, namely co-fusion of two different kinds of iron — one low in carbon and the other a high-carbon steel or cast iron. We present a suitable extract from Dr. Voysey's manuscript[23] describing the process:

> Konasamudram, situated about 12 miles south of the Godavari and 25 from Nirmal (mentioned in *Āin-ī-Akbarī*) is celebrated for its manufacture of steel, the chief part of which goes to Persia; the following are the chief details of the place, and personal inspection of the process.
>
> The furnace is a temporary circular structure of clay, from four to five feet in height, and five feet in diameter. It is sunk two feet below the surface of the ground. The fuel is charcoal. The granitic clay of the furnace is infusible. In making the pine-shaped crucibles, the same granitic clay, containing some quartz and felspar, is ground to a fine powder along with the fragments of old furnaces and crucibles, and the whole kneaded together with the chaff of rice and oil.
>
> No charcoal is put into the crucible, but small pieces of *kanch*, or the glass formed in the process, are put at the bottom as flux. The materials used in the preparation of the steel are two different kinds of iron: one from Mirtpalli — the other from Kondapur, in the proportion of three parts of

the former to two of the latter. The Mirtpalli iron, derived from iron sand (magnetite), is sent in the state of large amorphous masses of a reddish grey colour, extremely porous in texture and having iridiscent internal fracture. The Kondapur iron is procured from an ore found amongst the iron clay (laterite), at a place about 20 miles distant. It is said to be of a dirty brown colour, and very frangible. The iron is however moderately compact of a brilliant white fracture. Occasionally it contains some ingredient which spoils the steel, rendering it excessively brittle; the natives assert that the adulteration is copper, but it is more probably arsenic.

The mixture being put into the crucible, the fire is excited and kept up for 24 hours. The degree of heat, reckoned at 130° of Wedgwood, could fuse steel in 3 hours. The bellows made of bullock skins are plied incessantly day and night. During the operation the men are relieved every four hours, each being engaged to work 12 out of the 24. They are partly protected by a screen of mud between them and the furnace, but the heat and exertion render their task sufficiently laborious. After the operations, the crucibles are taken out and placed on the ground to cool.

When quite cold, a crucible is opened, and a cake of steel of great hardness is found, weighing on an average about a pound and a half. The cake is covered with clay, and annealed in the furnace for 12-16 hours. It is then taken out and cooled, and again annealed, and this may be repeated a third or fourth time until the metal is rendered sufficiently soft to be worked. The steel is known by the name of wootz.

The export of this metal to Persia must be profitable. We found at the village, in 1820, Hājī Hosyn (Hussain?) from Ispahan, engaged in the speculation, here again in 1823. He informed us that the place and the process are both familiar to the Persians, and that they have attempted to imitate the latter without success.[23]

Recent investigations by Lowe

Following Bronson's intelligent interpretation[7] of Voysey's report,[23] additional knowledge on the Hyderābād or Deccani process of the production of wootz has been gained through the recent investigations by Thelma Larson Lowe.[73-76] She has identified in her extensive field-work 74 sites of ore smelting, of which 14 were crucible production centres, 8 sites of historical mining activity and 17 probable ore sources.

The map (*fig*.4.1) shows the area of Andhra Pradesh bounded by 78° to 79° E and 18° 10' to 19° 10' N providing two types of iron ore. South of Nizāmābād there were laterite-smelting sites exploiting lateritic hardcaps around Inparattigutta. Shafts less then one meter in diameter and perhaps more than 4-5 meters deep were excavated. A transect walked across the approximate one-half kilometer width passes near 44 shafts. The debris around the shafts consisted of the redeposited alkalis leached from the siliceous rock-cap, and this could have served as a flux.

East of Nizāmābād and south of the Godāvarī river lay the magnetite banded iron formation (BIF) sites around Regunta-Rangaraopet-Konasamudram. Lowe noted[73] shallow cups carved into the surface of fine-grained granite which concentrated the heavier fraction of crushed magnetite-quartzite. A slurry of crushed ore was flushed over the cupped surface creating an uneven, swirling flow gradient that removed the lighter fraction, the quartz and other gangue minerals, and the magnetite accumulated in the cups. Havart recorded[63] the practice of roasting magnetite ore in Andhra Pradesh.

Lowe specially studied the Deccani crucibles[74-76] consisting of refractory vessel to hold the high-carbon iron charge and a cover of an appropriate geometry. The size of the crucible varied from 2.5 cm to 12 cm interior diameter. The Voysey account stated that for making the crucibles, charred (coked) rice hulls were mixed with clay and oil. It is known that during ignition, the surfaces of the polysilicic form of silica become partially dehydrated and are more compatible with oils than with water. This specific property of the surface of charred rice hulls was apparently exploited by Deccani refractory ceramicists.[74]

The charged crucible was fired at high temperature (1150-1250°C) in a strongly reducing atmosphere. Lowe *et al*. found through optical microscopy, XRD, SEM, TEM, HREM techniques that in the used crucibles there were two phases reinforcing the glassy matrix — cristobalite relics of rice husks and a network of acicular alumino-silicate mullite crystals.[76]

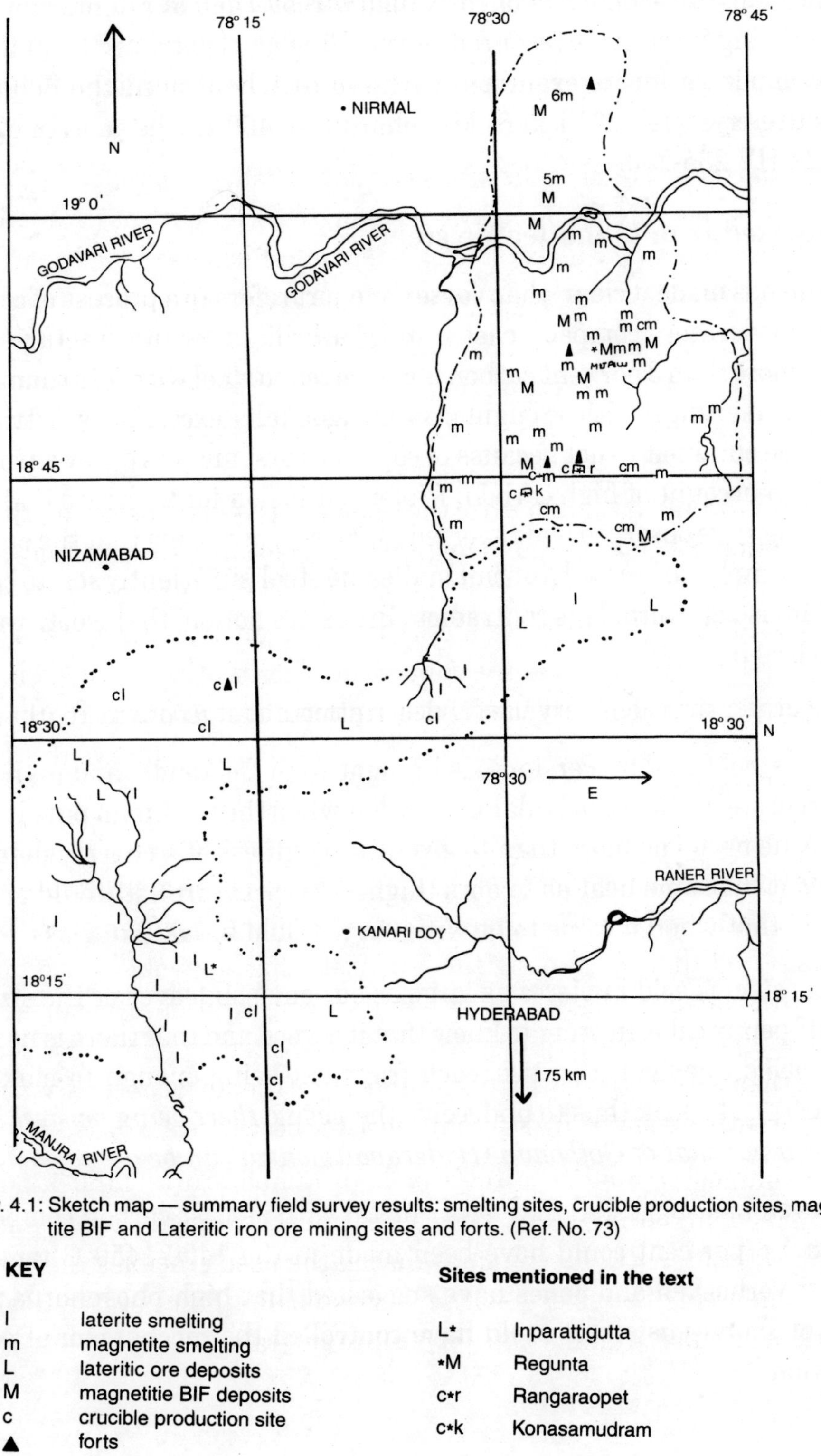

Fig. 4.1: Sketch map — summary field survey results: smelting sites, crucible production sites, magnetite BIF and Lateritic iron ore mining sites and forts. (Ref. No. 73)

KEY

l	laterite smelting
m	magnetite smelting
L	lateritic ore deposits
M	magnetitie BIF deposits
c	crucible production site
▲	forts

Sites mentioned in the text

L*	Inparattigutta
*M	Regunta
c*r	Rangaraopet
c*k	Konasamudram

Thermal properties controlled by high porosity and by orientation of the rice husks during fabrication favoured slow cooling and controlled crystallisation.[75] Konasamudram ingots examined by Lowe clearly showed the dendritic lath cementite crystals[75] of Vickers Microhardness 450 in the matrix of lamellar pearlite HV 235-250.

Carburised iron to malleable steel

Bronson has made it clear[7] that Voysey's report refers to a process of co-fusion in which 'moderately compact' cast iron 'of a brilliant white fracture' probably having more than 3 per cent carbon is converted to steel with less than 2 per cent carbon by melting it with wrought iron. Occasionally 'excessively brittle' product used to be obtained — not because of copper or arsenic, as Voysey surmised, but simply on account of high carbon. Prolonged fusion for 24 hours and repeated annealing were observed and reported by Voysey indicating that the end product was not simply carburised iron but malleable steel 'sufficiently soft to be worked' like Damascus sword. This contradicts Bronson's notion[7] that wootz was merely carburised iron.

Tavernier provided very useful description about *wootz* in 1679:

> This steel (used by Persians) is brought from Golkonda, and is the only sort of steel which can be damasked. For when the workman puts it in the fire, he needs no more than to give it the redness of a cherry; should he give it the same heat as to ours (higher temperature), it would grow so hard that when it came to be wrought it would break like glass.
>
> This steel is sold in pieces as large as our one-sou loaves or the size of a half-penny cake. In order to know that it is good and that there is no fraud involved, they cut it in two, each fragment being enough to make one saber. . . . I speak thus to undeceive the people *there being no steel in the world but that of Golkonda (Hyderabad) which can be damasked.*

Verhoeven has suggested that *wootz* metal with as-cast carbon composition of around 1.6 per cent could have been made in the 1400-1450°C temperature range.[77] Verhoeven and Jones have suggested that high phosphorus impurity levels of Damascus steels could have controlled the mechanism of cementite formation.[78]

Sherby and Wadsworth studied[79] how Damascus swords containing more carbon than most modern steel (*vide* Table 4.4) could have been prepared from Indian *wootz*. High-carbon *wootz* cakes were probably forged at a not-too-high temperature in the range of 650-850°C (between 'blood-red' 650° to 'cherry-red' 850°C) and then heat-treated by quenching.

Table 4.4

Composition of wootz steel and damascus blades[33]

	C %	*Si* %	*Mn* %	*P* %	*S* %	*Others* %	*Ref.*
1. Wootz (Royal Sch. of Mines, London)	1.34	-	-	-	-		78
2. Wootz (Sri Lanka)	1.97	0.007	0.007	0.002	0.007		70
3. Wootz, Indian	1.642	0.045	-	-	0.181		7
4. Wootz, Indian	1.682	0.042	-	-	0.17		7
5. Wootz, Mysore	0.963	0.127	0.097	0.007	0.02		7
6. Wootz, Mysore	0.45	0.14	-	0.27	0.01		7
7. Damascus blade	1.49	0.005	0.08	0.1	0.05		78
8. Dagger	1.677	0.015	0.056	0.086	0.007		78
9. Dagger	1.575	0.011	0.3	0.104	0.018		78
10. Sword	1.874	0.049	0.005	0.127	0.013		78
11. Sword	0.596	0.119	0.159	0.252	0.32		78
12. Sword	1.342	0.062	0.019	0.108	0.008		78
13. Sword	1.726	0.062	0.028	0.172	0.02		78
14. Sword	1.62	0.027	trace	0.087	0.007	Ni, Cr traces	78
15. Sword	1.42	0.11	0.13	0.035	0.038		78
16. Sword	1.42	-	0.015	0.206		-Ni:0.016; Cu:0.056	78
17. Sword	1.42	-	0.14	0.05		-Ni:0.008; Cu:0.041	78
18. Sword	1.1	-	0.071			-As:0.005; Cu:0.034	78

The European smiths failed to reproduce the results, probably because they were familiar with only low-carbon steel having a higher m.p., and tried to forge *wootz* at a white heat, when the cementite phase dissolved in the austenite, and the brittle structure 'crumbled under the hammer'.

Sherby and Wadsworth have shown[79] that high-carbon (1.3-1.9 per cent) *wootz* can be satisfactorily forged up to a temperature of 850°C without producing brittleness; slow cooling allowed a coarse, continuous and white cementite network which could be seen after selective etching, preferentially attacking the ferrite background. These damask patterns of cementite network have been poetically described as 'ten millions of (white) meandering lines in the dull blue sword'.

Heating and quenching produced martensitic structure providing additional hardness. If the Damascus blade were heated to *only just above* 727°C before being quenched, it would have been hard as well as tough (non-brittle). On the other hand, heating to much above 800°C before quenching would endow brittleness to the sword, thereby giving a bad reputation to Indian wootz!

Although the Nirmal-Konasamudram area in Andhra Pradesh was the most famous steel-making site in pre-modern India, there were other places where such art prospered. At Tendukhera in the Narasinghpur District, Madhya Pradesh, *pakkā* iron or crude steel used to be made and converted to edge tools by case hardening, during the middle of the nineteenth century. Before 1827, hot carburising and quenching gave steel in the iron works of Sāgar and Bundelkhaṇḍ. During this period, Cutch steel was famous. Abbott described in 1847 how cast steel (*wootz* type of material) was being converted in Gujarat to sufficiently elastic (Damascus type) sword blades.[80] The edge of the damask was not sufficiently malleable but 'seemed specially suited to the fabric of razors, penknives and surgical instruments, in which keenness of edge is of the first consequence and elasticity of none'.[80]

Different types of enterprises

Bhattacharyya identified three distinctly different types of economic organisation in the pre-modern iron and steel industries of India.[30]

TYPE I: TRIBAL HOUSEHOLD INDUSTRY

The Agāriās in the Central Provinces, the states of Rewā, Sargujā, Udaipur and Jaśpur, in the districts of Mirzāpur, Uttar Pradesh, Rāñchī and Pālamau in Bihar, etc., were iron-smelting tribes. In their smelting shop all labours were supplied by members of one family. The produce often passed to the *lohār*s or the ironsmiths of the nearby villages. The Agāriās in the inaccessible regions had to market their produce through itinerant merchants.

TYPE II: MARKET-ORIENTED CRAFT GROUPS

Skilled, free, specialist craftsmen operated centering local markets. In the district of Birbhūm, West Bengal, there was a flourishing iron-smelting industry centering around Deocha.[53] The iron-smelters purchased ore in established market places called *aurangs* from *beparries,* itinerant merchants. After smelting in the *koṭśāl*s, the Muslim smelters used to sell the iron blooms to the *kāmārśāl*s or the refineries where the refiners were from Hindu artisan castes.

Sanyal further subdivides[54] the manufacturing scenario at Birbhūm into two subtypes: IIa and IIb. The first sub-type of manufacture was practised in the western-most part of Birbhūm by the tribals: Santāls and Muṇḍa Kols. The tribals were in the habit of migrating from place to place in search of spots where the timber and ore supply were readily available. This tradition of itinerant smelters was mentioned in the *R̥gveda* and has always been a part of the Indian heritage. In Maharashtra, Dhavads, a wandering Muslim class, specialised in smelting.[81] The furnaces used by the tribals were small in dimension and period of service.

The second sub-type IIb of manufacture was practised by the settled Bengālī-speaking population concentrated in the north-western part of the district who had permanent *koṭśāl*s and *kāmārśāl*s as mentioned before. Their furnaces were much bigger, and iron formed at the bottom was in a *molten condition*. The refining was a sort of puddling process. The itinerant merchants carried the finished product, namely refined iron, to the accessory industries in Dubrājpur, Lokpur, Rājnagar, Rāmpurhāṭ, Murśidābād, etc., where knives, swords, daggers, axes, scissors, cooking utensils, etc., were made.

Production for the market was not as highly developed (as in Birbhūm) in places remote from the centres of commerce. In the Salem district of Tamil Nadu, wrought iron, made in Catalan type furnace and high quality steel or *wootz*, made in crucibles, were sold in the local bazaar. Similarly, the smelters in the Jorhāṭ district of Assam, supervised by an *ozhā* or master craftsman, produced lumps of iron and offered their produce for sale in the local bazaar.

TYPE III: PROTOCAPITALIST ENTERPRISES

In the first decade of the nineteenth century, proto-capitalist enterprises were already in existence in Mysore. Buchanan-Hamilton described the iron-smelting workshops in Mysore owned by proprietors who employed labourers, foremen and differentiated their wages according to their skills.[22] In certain places such as Madhugiri, Devarāyadurga, etc., money wages were not paid, but the physical produce was divided. In a typical case, ordinary workmen (numbering 20) received 3 per cent each, the foreman of the smelting house 6 per cent, the foreman of the forging house 8 per cent and the reminder 26 per cent went to the proprietor. From the sale proceeds of his share, the proprietor met the expenses of the 'firm' such as customs duties, forest duties for charcoal smelting, payments to the village headman for permission to gather ore, furnace rent to the latter, charity for the brāhmaṇas, etc.

Steel-making enterprises often received capital advances from merchants who would themselves borrow from'some greater merchant'. Crucible steel of Mysore and Hyderābād was in great demand and 'almost the whole' was exported. In 1832 Voysey met in Konasamudram an Ispahani merchant who was in the business of exporting steel to Persia.

Capt. L. Jacob observed a similar structure of enterprise in Kāṭhiāwār (Saurāṣṭra) in 1843. The proprietor employed about ten workmen in each smelting house, the collector of charcoal and ore and the master workman; he also paid taxes and material costs.

Certain features of the type of enterprise in Mysore, Hyderābād and Kāṭhiāwār may be summarised as follows: (a) higher level of physical output compared to type I; (b) a higher level of technology as evidenced by the production of crucible steel; (c) the proprietor in the role of a manager; (d) differentiation in functions and wages of labourers.

Stagnation to extinction: reasons

In many ways, the iron and steel industries of India were leading in the ancient, medieval and even pre-modern world. The outstanding achievements were made in the areas of carburisation of wrought iron and crucible steel making. Then why did the indigenous traditions stagnate and ultimately get extinguished?

There were two sets of factors which would explain the decline: (a) technical and (b) socio-political and intellectual. The second set was deeper in origin and more profound in results, affecting not only ferrous but also many other indigenous industries, in brief, science and technology in India as a whole.

The first set viz., technical factors were related mainly to the charcoal issue which we have deliberated upon earlier. We have emphasised that new technology had to be introduced all at once. Replacement of charcoal by coke was not the only issue. Even the British entrepreneurs failed, when they tried to introduce improvements in a piece-meal fashion. Thus, their criticism against the artisans' 'resistance towards technological improvements' was uncharitable.[30]

The second set of factors may be listed as far as possible in a chronological manner.

(1) The Hindu-Muslim rivalry prevented free diffusion of technical knowledge to and from India.

(2) Neither the Moghul government nor the people in the country were awake to the new technological developments in the field of mass production of iron and steel, which were spreading fast in Europe. They were very slow in innovation and mechanisation.

(3) In Britain, the demand for cast iron cannon led to the sudden increase in size and productivity of blast furnaces from 30 to 300 tons during the course of the fifteenth century. When the industry was rapidly burning up forests for fuel, the king intervened.[82] No such development took place in India on account of closed economy and lack of maritime adventure.

(4) During the Moghul period, guns were made mostly of bronze and brass. Cast iron cannon was never an article of urgent need.[61] The barrels of Akbar's guns were made of discs joined together by forge-welding. The expertise in forging techniques was therefore diluting the need for casting

technology.[61,83]

(5) Since the middle of the seventeenth century, the Marāṭhās purchased cannons from the Europeans. The casting technology for iron was never learnt. By the close of seventeenth century, the European experts grew reticent about disclosing technological information due to the fear of social ostracisation as well as legal action which was becoming increasingly evident in the West.[84] Later, this desire to isolate technology from the Indians became a part of the official colonial policy.

(6) The Britishers, who initially arrived as a trading company, started flooding the Indian market with cheap iron produced in Britain on a large scale. The rising cost of production by the indigenous methods forced the artisan families to quit their jobs for survival.

(7) The Britishers were very selective in extending the financial assistance to entrepreneurs interested in reviving Indian iron and steel industry. The proposal of Mr. Inder Narain Sharma for large scale production of iron in Birbhūm in 1774 was turned down, whereas a few British entrepreneurs were granted the permission.[53]

(8) During the latter part of the eighteenth century, the number of iron and steel furnaces in India was in the region of 10,000, producing on an average 20 tons per year per furnace. From about AD 1800 onwards, India was treated as a consumer of British manufactures, and Dharampal has rightly commented that the disappearance of the indigenous industries resulted from 'large-scale economic breakdown and hostile state policy'.[27]

(9) However, Dharmapal overshot his point when he asserted that the sciences and technologies of India 'had reached a desirable balance before the eighteenth-century' and 'were usefully performing the tasks desired by Indian society'.[27] We wish we could accept his assertion.

Dharampal referred to the 'link' (?) between the practitioners of the various techniques and 'the professors of the theoretical knowledge'. The link, 'largely snapped in India by the end of the eighteenth-century', could have been re-established, according to Dharampal, 'in a changed political climate'. These are indeed brave patriotic words!

We are not aware of any Indian 'professor of the theoretical knowledge' in the pre-modern era who could explain the extractive and physical metallurgy of iron and steel. As a matter of fact, the theoretical knowledge pertaining to chemistry and metallurgy was barely evolving at that time even in the West, and the Japanese people could easily see that this new knowledge had to be borrowed from the West.

When Dharampal opines that 'unthinking transplanting of European sciences and technologies in India resulted in retarding and blunting of indigenous innovation and creativity',[27] the topic certainly demands detailed discussion and debate. Of course there is a merit in systematic studies on the ancient technologies. The pre-modern iron and steel industries of India held certain promises which can be considered for exploitation in the modern context. Rice-husk reinforced refractory crucibles can still be tried for high temperature work. The ancient art of making wootz steel and Damascus blade can be simulated to make ultra-high carbon steel having strength as well as ductility.[79] Some scholars believe that the pre-modern techniques of direct smelting of iron ore in small bloomeries and of crucible production of special steels may be revived with certain modifications.[33]

References

1. A.K. Biswas, *Minerals and Metals in Ancient India*, Indian National Science Academy Project (1987-1991) Report, Volume I. *Archaeological Evidences*. Chapters XII entitled 'Iron in Ancient India' and XIX entitled 'Iron in the Ancient India of the Post-Christian Era'. The book published by D.K. Printworld (P) Ltd, 1996.
2. O.P. Agrawal, Hari Narain, Jai Prakash and S.K. Bhatia, Lamination Techniques in Iron Artifacts in Ancient India, *Historical Metallurgy*, 24, (1) 1990, pp. 12-26.
3. N.R. Banerjee, *The Iron Age in India*, Munshiram Manoharlal, Delhi, 1965.
4. D.K. Chakrabarti, The Issues of the Indian Iron Age, in *Recent Advances in Indian Archaeology* (1983 Poona Seminar), ed. by S.B. Deo and K. Paddayya, Deccan College, Poona, 1985, pp. 74-88;

   (b) *The Early Use of Iron in India*, Oxford University Press, Delhi, 1992.
5. K.T.M. Hegde, A Model for Understanding Ancient Indian Iron Metallurgy, *Man.*, 8, (3) September 1973, pp. 416-21.
6. B. Prakash and V. Tripathi, Iron Technology in Ancient India, *Historical Metallurgy*, September 1986, pp. 568-79; (b) B. Prakash, Metallurgy of Iron and Steel Making and Blacksmithy in Ancient India, *Indian J. History of Science*, 26, (4), 1991, pp. 351-71.
7. Bennet Bronson, The Making and Selling of Wootz, A Crucible Steel of India, *Archaeomaterials*, 1, 1986, pp. 13-51; Comments by B. Prakash and reply by B. Bronson,

3, 1989, pp. 95-6.

8. Tapan Raychaudhuri, and Irfan Habib (ed.), *The Cambridge Economic History of India*, Vol. I (AD 1200-1750), Cambridge University Press and Orient Longman, New Delhi, 1993.
9. S.D. Goitein, *Studies in Islamic History and Institution*, Leiden, 1966, p. 339 quoted in ref. no. 8.
10. Fakhr-i-Mudabbir, Ādābu'l — Harb wa'sh — Shujāat, ed. by Ahmad Suhaili Khvānsarī, Tehran, A.H.S. 1346, quoted in ref. no. 8.
11. Ref. no. 8, p. 275, 310 and 322.
12. W.H. Moreland, *From Akbar to Aurangzeb — A Study in Indian Economic History*, Oriental Book Reprint Corporation, Delhi, 1972, p. 192.
13. H.K. Sherwani, "Golkonda" in *Encyclopaedia of Islam*, Vol. 2, pp. 1118-19.
14. Ref. no. 8, p. 323.
15. *Ibid.*, pp. 491-4.
16. S.D. Joshi, *History of Metal Founding on the Indian Sub-Continent Since Ancient Times*, Ranchi, 1970, pp. 58-86.
17. H. Blochmann (tr.), The *Āin-ī-Akbarī* of Abul Fazl, Oriental Books Reprint Corporation, New Delhi, 1977, Vol.1, p. 34 also footnote, pp. 115-22, p. 285 and footnote.
18. I.G. Khan, Metallurgy in Medieval India Sixteenth to Eighteenth Centuries, in *Technology in Ancient & Medieval India*, ed. by Aniruddha Roy and S.K. Bagchi, Sundeep Prakashan, Delhi, 1986, pp. 74-6.
19. Valentine Ball, *A Manual of the Geology of India*, Part III, *Economic Geology*, Geological Survey of India, 1881, pp. 335-414.
20. Captain Alexander Hamilton, Account of a Voyage to the East Indies, 1708, Vol. I, p. 395 quoted by S.G.T. Heatly, *Journal of the Asiatic Society of Bengal*, Vol. 12, 1843, pp. 542-3.
21. Dr. Benjamin Heyne, Iron Works at Ramanakapettah, report sent to Governor of Madras September 1, 1795, reprinted in Reference No. 27, pp. 263-7. Also included in Heyne's Tracts, *Historical and Statistical on India, with Journey through Various Parts of the Peninsula*, London, 1814.
22. F. Hamilton-Buchanan, *A Journey from Madras through the Countries of Mysore, Canara and Malabar*, London, 1807.
23. H.W. Voysey, Description of Native Manufactures of Steel in South India, *Journal Asiatic Society of Bengal*, I, 1832, pp. 245-7, II, p. 402.
24. W. Cracroft, Smelting of Iron in the Kasia Hills, *Ibid.*, I, 1832, pp. 150-1.
25. Major James Franklin, Observations on Several Iron Mines in Central Part of India, 19th May, 1835, India Office Library, Reprinted in Reference No. 27, pp. 269-94.
26. H. Yule, Notes on the Iron of Kasia Hills, *JASB*, XI, 1842, pp. 853-7; *vide* also Oldham, *Mem. GSI*, Vol. I, p. 201.
27. Dharampal, *Indian Science and Technology in the Eighteenth Century*, Academy of Gandhian Studies, Hyderabad, 1983.

28. John Percy, *Metallurgy of Iron and Steel*, London, 1864, p. 254.
29. B. Prakash, Methods of Iron-Making in Early India, International Symposium on "The Archaeometallurgy of Iron", Liblice, Czechoslovakia, October 1987.
30. Sabyasachi Bhattacharyya, Iron Smelters and the Indigenous Iron & Steel Industry of India: From Stagnation to Atrophy, *J. Indian Anthrop. Soc.*, 5, 1970, pp. 133-51.
31. H.C. Bhardwaj, Development of Iron and Steel Technology in India During Eighteenth and Nineteenth Centuries, *Indian Journal of History of Science*, 17 (2), 1982, pp. 223-33.
32. K.K. Prasad, *et al.*, Attempt on Revival of Rural Iron-making, *Metal News*, 12, 1, 1990, p. 1.
33. "Indigenous Iron and Steel Technology of India", Monograph Prepared by A.V. Balasubramanian, N.B. Ballal, Bhanu Prakash *et al.* for the Congress on Traditional Sciences and Technologies of India (CTSTI), Nov. 28-Dec. 3, 1993, I.I.T. Bombay, 114 pages.
34. V.H. Dirar, K. Aravindakshan, *et al.*, Studies on Iron-smelting Practice at Malabar, Paper Presented in the CTSTI, I.I.T. Bombay, 1993.
35. C. Vivek, Experiments on Indigenous Iron Smelting, B.Tech. Dissertation, 1993, I.I.T. Bombay.
36. J. Campbell, Assistant Surveyor-General, Madras Establishment, Public Consultations, 16th August, 1842, pp. 4160-84, at the Madras Records Office.
37. A. Rahman, and B.V. Subbarayappa, A Note on the Native Method of Bar Iron Production in South India (Salem Region), *Indian J. History of Science*, 1, (2), 1967, pp. 158-61.
38. Ref. No. 27, pp. 295-307.
39. Ref. No. 19, p. 356.
40. Ref. No. 27, pp. 263-7.
41. Ref. No. 27, pp. 269-94; Ref. No. 19, pp. 385-7.
42. Ref. No. 19, p. 387.
43. Pramatha Nath Bose, The Iron Industry of the Western Portion of the District of Raipur, *Records of the Geological Survey of India*, Vol. XX, Part 4, pp. 167-70.
44. Verier Elwin, *The Agaria*, Oxford University Press, Calcutta, 1942.
45. M.S. Krishnan, *Iron Ores of India*, Indian Association for the Cultivation of Science, Calcutta, 1955.
46. B. Prakash, and K. Igaki, Ancient Iron-Making in Bastar District, *Indian Journal of History of Science*, 19 (2), 1984, pp. 172-85.
47. M. Ghosh, Primitive Iron-Making in Contemporary India, *TISCO Journal*, 11, (3), 1964, pp. 132-4.
48. Ref. No. 19, pp. 412-14.
49. A.K. Biswas, Non-Gem Minerals in Pre-modern India, *Indian Journal of History of Science*, 29(3), 1994, pp. 421-63.
50. F.R. Mallet, On Some Forms of Blowing-Machine Used by the Smiths of Upper Assam, *Records of the Geological Survey of India*, Vol. X, Part 3, pp. 152-7.

51. Alexander Hamilton, *Account of a Voyage to the East Indies*, Vol. 1, 1708, pp. 392-5, quoted in Ref. Nos. 52 and 53.

52. Ref. No. 19, p. 361, Ball Comments:

"It was not stated by whom the process of working in cast iron (at Balasore) was introduced, but there were at that time factories belonging to the English, Dutch and French."

53. S.G.T. Heatly, Contributions Towards a History of the Development of the Mineral Resources of India, *Journal of Asiatic Society of Bengal*, Vol. XII, 1843, pp. 542-6.

54. Hitesranjan Sanyal, The Indigenous Iron Industry of Birbhum, *Indian Economic & Social History Review*, Vol. 5, No. 1, 1968, pp. 101-8.

55. Ref. No. 19, pp. 362-4.

56. Ref. No. 19, pp. 376-81.

57. Ref. No. 19, pp. 367-8 quoting *Proc. Asiatic Society of Bengal*, August 1880.

58. A.E., *Iron Works at Firozpur, Gleanings of Science*, Vol. III October 1831, pp. 327-8.

59. Ref. No. 19, pp. 400-2 quoting Captain Jacob's description. Also Report on the Iron of Kathywar, *Journal Royal Asiatic Society*, Vol. VII, 1843, pp. 101-2.

60. S. Bloomgren, and E. Tholander, *Scandinavian Journal of Metallurgy*, 15, 1986, p. 151.

61. I.G. Khan, Metallurgy in Medieval India Sixteenth to Eighteenth Centuries, in *Technology in Ancient and Medieval India*, ed. by Aniruddha Roy and S.K. Bagchi, Sundeep Prakashan, Delhi, 1986, pp. 71-91.

62. D.A. Kulkarni, *Rasa-Ratna-Samuccaya*, Meherchand and Lakshmandas, Delhi, 1969; the text and translation of the relevant portion (5.69-5.96) available in *Indian Journal of History of Science*, 24 (2), April 1989, pp. 217-20 also *ibid.*, 26 (4), October 1991, pp. 354-5 discussion in ref. no. 6b.

63. D. Havart, *Rise and Decline of Coromandel*, Original Dutch in 1692 from Utrecht, English version quoted by Dharampal[27].

64. George Pearson, Experiments and Observations to Investigate a Kind of Steel, *Philosophical Transactions*, Vol. 85, 1795, pp. 322-46.

65. D. Mushet, Experiments on Wootz, *Proc. Royal Soc.*, I, 1805, pp. 183-5.

66. J. Stodart, A Brief Account of Wootz, *Asiatic Journal*, V, 1818, p. 570. Stodart assisted Michael Faraday in analysing Indian wootz.

67. Michael Faraday, An Analysis of Wootz or Indian Steel, *Quarterly Journal of Science, Literature, and the Arts*, Vol. 7, 1819, pp. 319-30.

68. J. Stodart, and M. Faraday, Experiments on the Alloys of Steel, Made with a view to its Improvement, *Ibid.*, Vol. 9, 1820, pp. 319-30.

69. J.M. Heath, On Indian Iron and Steel, *Journal Royal Asiatic Scoiety*, Vol. 5, 1839, pp. 390-7.

70. A.K. Coomaraswamy, *Mediaeval Sinhaless Art*, Chapter XI, New York, 1908, 1956 edition.

71. C.V.B., Method of Making Steel in Mysore, *Asiatic Journal*, Vol. 9, 1820, p. 441.

72. K.N.P. Rao, J.K. Mukherjee, and A.K. Lahiri, Some Observations on the Structure of Ancient Steel from south India and its Mode of Production, *Bulletin of the Historical Metallurgy*, Vol.4, No. 1, 1970, pp. 12-17.
73. Thelma Larson Lowe, Indian Iron Ores and the Technology of Deccani Wootz Production, Iron Paleometallurgy and Cultures, International Symposium, Belfort, November 1990.
74. T.L. Lowe, Refractories in High-Carbon Iron Processing: Deccani Wootz-Making Crucible *in* Ceramics and Civilisation, Conference, April 1987, American Ceramics Society, Pittsburgh.
75. T.L. Lowe, Solidification and Crucible Processing of Deccani Ancient Steel, *in* Solidification and Materials Processing, Conference Procceedings, DMRL, Andhra Pradesh, January 1988, IBH and Oxford Press, New Delhi.
76. T.L. Lowe, N. Merk, and G. Thomas, An Historical Mullite Fiber-Reinforced Ceramic Composite: Wootz Crucible, Symposium on Materials Issues in Art and Archaeology, San Francisco, April 1990, Proceedings Editor, Pamila Vandiver.
77. J.D. Verhoeven, Damascus Steel, Part I: Indian Wootz Steel, *Metallography*, Vol. 20, 1987, pp. 145-51.
78. J.D. Verhoeven, and L.L. Jones, Damascus Steel, Part II: Origin of the Damask Pattern, *Metallography*, Vol. 20, 1987, pp. 153-80.
79. O.D. Sherby, and J. Wadsworth, Damascus Steels, *Scientific American*, Feb. 1985, pp. 112-20.
80. Capt. James Abbott, Process of Working the Damascus Blade of Gujarat, *Journal of the Asiatic Society*, Vol. 16, May 1847, pp. 417-23; *ibid.*, Vol. 16, July 1847, pp. 666-7.
81. Dharma Kumar, *The Cambridge Economic History of India*, Vol. II (AD 1757-1970), Cambridge University Press and Orient Longman, 1982, p. 349.
82. R.F. Tylecote, *Metallurgy in Archaeology*, Arnold Publishers, London, 1962, pp. 296-7.
83. I.G. Khan, Metallurgy in Medieval India — The Case of Iron Cannon, *Proceedings of Indian History Congress 1984*, University of Delhi, Published 1985, pp. 464-71.
84. A.J. Qaisar, *Indian Response to European Technology and Culture*, Oxford University Press, 1982, pp. 47-50, 145-6.

5

Brass and Zinc in Pre-Modern India*

THE earlier occurrence of zinc in man-made artifacts is in the form of the copper alloy known as brass. Ever since the discovery of copper and the alloying elements of tin, arsenic, lead, etc., different materials, including zinc, were used to alloy and harden copper. India's primacy in brass and zinc-making has been documented.[1-26]

The earliest method of making brass was possibly the cementation process in which finely divided copper fragments were intimately mixed with roasted zinc ore (oxide) and reducing agent, such as charcoal, and heated to 1000°C in a sealed crucible. Zinc vapour formed dissolved into the copper fragments yielding a poor quality brass, zinc percentage of which could not be easily controlled.

Fusion of zinc with copper increases the strength, hardness and toughness of the latter. When the alloy is composed of 10-18 per cent zinc, it has a pleasing golden yellow colour. It can also take very high polish and literally glitter like gold. For this property, brass has been widely used for casting statuary, covering temple roofs, fabricating vessels, etc.

Reduction of zinc oxide around 1000°C is crucially important; below 950°C no zinc is produced. Zinc is obtained in vapour form at this temperature, since its b.p. is 913°C. With trace oxygen, the zinc vapour would be reoxidised and hence the successful operations in the past must have been done in closed crucibles. If the temperature were higher than 1083°C during brass-making, then copper would melt and flow down to the bottom of the crucible forming a puddle there, exposing a very small surface area of the metal for alloy formation.

---

* This material has been published with the title 'The Primacy of India in Ancient Brass and Zinc Metallurgy', in *The Indian Journal of History of Science 28*, (4), 1993, pp. 309-30.

Brasses containing up to 36 per cent zinc are known as α-brasses, which undergo easy cold work. Brasses containing more than 46 per cent zinc are brittle. With zinc content between 36 and 46 per cent, we have $\alpha + \beta$ brasses which are lighter, harder and more suitable for casting statuary.

Werner[27] and Haedecke[28] demonstrated experimentally that brass produced by the cementation process could not contain more than 28 per cent zinc. Brass founders trying the cementation process have verified this observation.

The materials of antiquity containing more than 28 per cent zinc in copper matrix must have been prepared by mixing the two metals, which could have been possible only after the discovery of zinc as a separate metal and its preparation by a process such as distillation. The antiquity of brass artifacts can, therefore, be divided into two eras, one preceding, and the other following the discovery of zinc as a separate metal.

Brass before the discovery of zinc

We claim that the earliest artifact noted so far containing an appreciable amount of zinc anywhere in the world is from India. Lothal (2200-1500 BC) showed one highly oxidised antiquity (No. 4189), which assayed: 70.7 per cent copper, 6.04 per cent zinc, 0.9 per cent Fe and 6.04 per cent acid-soluble component (probably carbonate, a product of atmospheric corrosion).[29] The material could have been prepared through smelting of zinc-bearing copper ore or the cementation route described earlier. The raw materials might have come from the Ahar-Zāwar area. The Harappan site of Rosdi, also in Gujarat, has yielded a few samples of chisel, celt, rod and bangle, made of brass and assaying up to 1.54 per cent zinc.[30]

Similar materials might have been used for making the brass-bronze items of Atrañjikhera during the PGW-era (1200-600 BC).[31] One copper-based item contained 11.68 per cent Sn, 9.0 per cent Pb and 6.28 per cent Zn, while another item assayed 20.72 per cent Sn and 16.20 per cent Zn. Both the samples contained traces of iron and sulphur, indicating the possibility of chalcopyrite and sphalerite-galena having been the source materials, which could easily come from the Ahar-Zāwar area. Most of the brass samples in ancient India contained variable proportions of Zn, Sn and Pb (Table 5.1).

We have drawn attention to the brass items of Lothal and Atrañjikhera and their possible link with the 1260 ± 160, 1136 ± 160 and 1050 ± 150 BC C-14 dates of the timber samples in the Rājpura-Daribā silver-lead-zinc mine near Udaipur.[4,20]

During the Harappan era, copper used to be alloyed with tin and arsenic;

Table 5.1

Analysis of some brass objects in ancient India

Sl. No.	*Date and site*	*Object*	*Analysis (%)* Cu	Zn	Sn	Pb	*Remarks*
1.	~ 1500 BC Lothal	No. 4189 Copper object	70.70	6.04	—	—	Ref. 29
2.	Harappan, Rosdi	Chisel, Celt rod, bangle	95.5-98.5	up to 1.54	up to 1.23	up to 1.20	Ref. 30, pp. 36-7
3.	~ 1000 BC Atrañjikhera	No. 1 No.2	70.99 52.42	6.28 16.20	11.68 20.72	9.00 9.77	Ref. 31
4.	4th century BC, Taxila	Vase Bm 215-284	55.39	34.34	4.25	3.08	Ref. 32
5.	2nd century BC Taxila	Bangle	73.72	19.70	0.10	5.84	Ref. 32
6.	2nd century AD Gujarat	Female figure carrying flower-container Indo-Parthian	88.60	7.60	0.13	2.49	Ref. 30, pp. 56-7
7.	5th century AD	Gāndhāra Buddha	68.50	20.25	3.86	3.62	Ref. 37
8.	6th century AD Akota	Ambikā	76.70	16.32	1.61	4.04	Ref. 30, pp. 104-5
9.	7th century AD Mahudi	R̥ṣabha-nātha	66.00	12.80	5.90	2.50	Ref. 30, p. 66
10.	8th century AD Kashmir	Śiva	82.00	17.00	1.00	—	Ref. 38
11.	9th century AD Nālandā	Buddha	78.95	15.15	0.74	3.03	Ref. 39
12.	11th century AD W. Tibet	Mañjuśri	65.50	30.40	0.30	1.70	Ref. 40

since these were scarce commodities, alternative alloying elements had to be looked for. Artisans in the Rajasthan-Gujarat region might have stumbled on to zinc ore deposit as a new source of alloying element.

Craddock *et al.*[6,7] surveyed the evidences of early brass artifacts in the West. The earliest brass artifacts known in the West come from excavations at the Gordion Tomb in Phrygia, dating from the eighth and seventh centuries BC onwards. These came after the Lothal and Atrañjikhera traditions. From the seventh century BC, the Greeks commented upon brass or oreichalkos, but always as an expensive, exotic metal not produced in Greece. There was no zinc in the early Greek bronzes. Etruscan bronze of the fifth century BC contained 11 per cent zinc.

Zinc metal and high-zinc brass

The earliest brass containing more than 28 per cent zinc, which could be made only after the isolation of pure zinc metal,[27,28] came from Taxila.[32] Craddock[8] pointed out the overriding importance of the vase (BM 215-84) excavated from the Bhir Mound at Taxila and dated to the fourth century BC. This brass sample contains 34.34 per cent zinc, 4.25 per cent Sn, 3.0 per cent Pb, 1.77 per cent Fe and 0.4 per cent nickel.[32] This is very strong evidence for the availability of metallic zinc in the fourth century BC. Possibly India was the first to make this metal zinc (*rasaka*) by distillation process, as practised for the other metal mercury (*rasa*).

There are references to zinc and brass in the lost (fourth century BC) text *Philippica* of Theopompus, quoted by Strabo in his *Geography*:

> There is a stone near Andreida (north-west Anatolia) which yields iron when burnt. After being treated in a furnace with a certain earth it yields droplets of false silver. This, added to copper, forms the so-called mixture, which some call *oreichalkos*.
>
> — Strabo, *Geography*, Book XIII, Sec. 56

The above reference pertains probably to the process of downward distillation of zinc ('droplets of false silver') and its subsequent mixing with copper to make brass *oreichalkos* (*ārakūṭa* in Kauṭilya's *Arthaśāstra*) described in details in the post-Christian era Sanskrit texts.

It is quite possible that the zinc-making technology travelled West from India during sixth-fifth centuries BC, as it did later again in the eighteenth-century AD. The pseudo-Aristotelian work *On Marvellous Things Heard* mentioned:

> They also say that amongst the Indians the bronze is so bright, clean and free from corrosion that it is indistinguishable from gold, but that amongst the cups of Darius there is a considerable number that could not be distinguished from gold or bronze except by colour.
>
> — quoted by Craddock[6]

The Indian emphasis was on the 'gold-like' brass and not on the zinc metal. The Greeks, however, used zinc metal as such in a few cases. In the course of the excavation of the Agora in Athens, a roll of sheet zinc was found in a sealed deposit dating from the third or second century BC.[33] Analysis showed it to be nearly pure zinc with 1.3 per cent lead, 0.06 per cent Cd, 0.016 per cent Fe and 0.005 per cent Cu with traces of Mn, Mg, Sn, Ag and Sb (quoted by Craddock).[6] Although Needham and Forbes doubted the above evidence on the ground that 'the pieces were beyond the contemporary technology',[8] Craddock certifies this to be a genuine sample.[8,24] It is quite possible that the Greeks had carried the material or the technology which had existed in Taxila as early as fourth century BC and possibly much earlier in Rajasthan.

Mining archaeology and smelting related to Indian zinc ore

The recent pioneering work on the zinc-lead-silver mining archaeology in the southern part of Rajasthan by Willies *et al.*[12,20,21] and the relevant C-14 dates have firmly established India's primacy in non-ferrous ore mining in the ancient world.

The ancient workings in the South Lode (100 m depth) of Rājpura-Daribā mine (80 km north-east of Udaipur) have been C-14 dated as 1260 BC, 1130 BC, 1050 BC and the East Lode workings as 375 BC, 360 BC, 250 BC, 120 BC, AD 150, etc. Thus, it is clear that the tradition of underground mining in India goes back to the thirteenth century BC, if not earlier. The earliest emphasis was possibly on copper ore; at Rājpura-Daribā, the other targets were lead, silver and possibly zinc ore, which is strongly suggested by the brass artifacts of Lothal and Atrañjikhera.

The art of smelting zinc ore and recovery of zinc metal by distillation must have been discovered before fourth century BC when Taxila produced the brass vase containing 33.34 per cent zinc.[32] This possibliliy is reinforced by the facts of mining archaeology. Starting from the fifth century BC, we have many C-14 datings in Rājpura-Daribā, Rāmpura-Aguchā (40 km south of Ajmer) and most crucially, in the Zāwar mine systems.

Zāwar (24° 21'N, 73° 41'E) is about 30 km south-west from Udaipur, where the ancient mines (earliest C-14 date obtained so far is 430 BC) are found, both opencast and underground. Zāwar Mālā, Mochiā and Balariā are some of the specific mines in this area.

The host rock of the Zāwar mines is sheared dolomite, the result of metamorphism of sedimentary dolomites. The ore was geologically deposited syngenetically as disseminated lenses within the dolomite beds. Zinc occurs as sphalerite or as marmatite in which zinc sulphide is in solid solution with iron sulphide. Also associated with the rock are galena, hydrozincite, pyrite, silver, etc.

Ancient workings are found at outcrops on the ridge of Zāwar Mālā, and as deep as the 470 m (above sea) levels of the modern Zāwar Mālā mine, some 120 m below surface. The upper parts accessible from there were a few tens of metres below surface, isolated by a roof fall. At the surface, mine openings occur at intervals of 50 m or so. Willies[20] investigated one mine of more commodious proportion — Pratāpkhān or Pratāp's mine — in which Rāṇā Pratāp Singh, rival of Akbar, took refuge during AD 1595-1600. A flat 'room' floored with phyllite slabs is inferred as Pratāp's refuge. The quarried materials used to feed a zinc smelter just below the narrow valley. (Vide Plates 23 & 28).

The earliest C-14 datings in the Zāwar mines are 430 ± 100 BC of the PRL 932 sample from the Zāwar Mālā mine and 380 ± 50 BC of the BM 2381 sample from the Mochiā mine.[17] Similar datings from Rājpura-Daribā (e.g., 375 BC), Rāmpura-Aguchā (370 BC), etc., confirm widespread underground mining of lead-zinc ores in the southern Rajasthan during the fifth-fourth centuries BC onwards.

Subsequent C-14 datings in the said mining area are 250 BC, 200, 170, 140, 120 BC, AD 60, AD 110, AD 150.

As regards the recovery of zinc from the ore, the crucible reduction/distillation method was put to large scale commercial practice in the thirteenth century AD; this will be described later. Indirect and circumstantial evidences suggest that distillation method was in vogue much earlier, probably from the fourth century BC onwards, although not on a large scale, as we find in the thirteenth century AD context.

In this connection, we recall the evidence from Rāmpura-Aguchā.[17,34] The zinc-lead-silver ore at the site was selectively mined at least as early as 370 and 250 BC. An appreciable amount of zinc must have been separated from the zinc-rich ore (present-day ore in the site contains 13.5 per cent zinc), as revealed from the low-zinc content slag. One sample of slag assayed as low as 0.01 per cent zinc.[34] Near the slag dump area several retort-like pieces were reported. When assembled, their appearance suggested a cylinder approximately 20 cm long with walls 4-5 cm thick and an innermost pipe-like feature with a coating of dirty white material, mainly zinc sulphate. They could be mistaken for tuyeres but for their closed pointed ends. This is highly suggestive of a used retort. Along with this, some thin-walled tube-like object containing a thin coating of blister type material was also found.[17]

It is conceivable that the retorts were being used in the said context for roasting zinc ore to obtain the light, white, smoky zinc oxide, which the ancient Greeks called pompholyx or philosopher's wool. In the modern zinc plant at Udaipur, roasting of zinc sulphide concentrate produces not only zinc oxide (and sulphur dioxide gas) but also some zinc sulphate, which was detected in the fourth-third century BC retort in Rāmpura-Aguchā.

The said retorts, already found sealed at one end, must have been closed or sealed at the other and also to prevent the escape of smoky zinc oxide into the atmosphere. The retorts were possibly modified to serve as reduction distillation chambers (to produce metallic zinc), the final version of which, notified in the thirteenth-century AD context, would be described later. Very significantly, Tiwari *et al.*[34] noted that the slag sample from Aguchā containing only 0.01 per cent zinc but as high as 9.30 per cent lead, was 'attached to baked earthen materials which could be part of the earthen appliance used for smelting'. We suggest the possibity that the earthen appliance was a zinc distillation retort. Remains of zinc furnaces have been found at Sojat in Jodhpur also.

Brass and zinc in ancient India

We may now turn our attention to the antiquity of brass in ancient India. Before the discovery of zinc metal in India (made by the distillation route) sometime during the fifth-fourth century BC, brass could be made*, as in Lothal and Atrañjikhera, only by the cementation route in which one of following was smelted along with copper ore: zinc ore, sphalerite concentrate or the roasted product, philosopher's wool or zinc oxide. The traditions of making philosopher's wool and cementation brass could have persisted even after the discovery of the distillation process of making zinc. We invite attention of the readers to the analysis of some Indian brass objects made before fourth century BC (Table 5.1).

As we have indicated earlier, the distillation route of making zinc and alloying this with molten copper was the only way of making high-zinc (more than 28 per cent) brass, such as the fourth-century BC Taxila vase (34.34 per cent zinc). The said vase (*BM* 215-84), excavated from the Bhir Mound site, was made before the Greek settlement in Sirkap.[32]

One bangle from the second-century BC Sirkap settlement assayed 19.70 per cent zinc. The Dharmarājika settlement of the post-Christian era produced brass objects like bangle and pot with controlled compositions 77-79 per cent Cu, 12.88-13.07 per cent Zn, 2.5-3.5 per cent Sn and 3-6 per cent Pb.

Some of the other early brass samples from ancient India have been reviewed by Neogi[35] and Ray,[36] an extract of which is presented below.

Brass articles of the first century BC or AD have been found on excavation of some ancient *stūpa*s. General Ventura executed operations for the examination of the *stūpa*s at Manikyalaya in 1830. Three deposits were obtained, of which the

---

* The second-century AD text *Rasaratnākara* of Nāgārjuna mentions (1.3): "What wonder is it that calamine (zinc ore) roasted thrice with copper converts the latter into gold." This is an allusion to the making of yellow gold-like cementation brass. The same text (1.31-2) also refers to the reduction-distillation of calamine yielding zinc: "an essence of the appearance of tin".

third, at a depth of 64 ft consisted of a copper box enclosing a brass cylindrical box cast and beautifully turned on the lathe. The lid of the brass casket was found on cleansing to be inscribed. From the inscriptions on the various articles of this deposit and the accompanying Indo-Scythian coins, the great tope at Manikyalaya has been identified to be a mausoleum of the Indo-Scythian king Kaniṣka (first century BC or AD).

Another inscribed brass urn of the same date as the former has been discovered in a tope about 30 miles west of Kabul in Wardak district. This urn, which in shape and size approaches closely the ordinary water-vessels in use in India to this day, was originally thickly gilt and its surface has in consequence remained well-preserved.

As regards coins, both brass and bronze were used in ancient India for coinage. Circular punch-marked brass coins of Dhanadeva and Āryavarmā of Ayodhyā (*c.* first century BC) have been found. Brass coins of kings of several other dynasties living at that time have also been collected. From these archaeological and numismatic evidences it is clear that brass was in common use in ancient India during the first century BC. A small number of die-struck coins of the Pre-Gupta and Gupta periods, including a piece attributed to Candragupta II, are considered to be made of brass.

Table 5.1 features some of the typical brass objects in ancient India up to eleventh century AD, before the advent of Muslims in the country. In his masterly treatise,[30] Swarnakamal highlighted the continuing role of Gujarat in brass technology ever since the Harappan era. The discovery of three important hoards of metallic art objects at Mahudi of north Gujarat, Lilvadeva (north-east) and Akota of central Gujarat, dated between sixth and eleventh centuries AD, proved that the artisans there had developed four varieties of alloys: (a) bronze, (b) zinc-bronze, (c) lead brass, and (d) conventional brass. Apart from Zāwar ore, the copper-zinc ore deposit of Ambāmātā (north Gujarat) was also used for smelting. A large number of brass coins of India dated fifth-twelfth centuries AD have been analysed by other workers.[37-40] A few of these are listed in Table 5.1. The data are impressive enough to justify the primacy of India in brass and zinc metallurgy.

Literary evidences on zinc and brass in ancient India

We have mentioned earlier the pseudo-Aristotelian work *On Marvellous Things Heard* referring to 'gold-like bronze' or brass of the Indians during the reign of Darius. The Greeks' 'oreichalcos' was equivalent to Kauṭilya's *ārakūṭa* (*Arthaśāstra*, II.12.23, II.17.14, IV.1.35). Both Kauṭilya and the earliest Indian brass of Taxila belonged to fourth century BC, while the earliest C-14 dating of the Zāwar mine is 430 ± 100 BC.

The technical term *ārakūṭa* for brass persisted through centuries and we find this mentioned in the fourth century AD Jaina text *Aṅgavijja* (as *hārakūḍa*) and also in *Amarakośa* (AD 450).

A more popular name for brass in the ancient India has been *rīti* or *rītikā*, which also meant calx of brass. The word was probably derived from *hārīta* or yellow, which had been a synonym for gold in the Vedic literature. The word was chosen on account of the yellow colour of gold-like brass. *Caraka-saṁhitā* refers to *rīti* (Sūtrasthāna, 5.26) and *rītikā* (Siddhisthāna, 3.4) amongst other metals and alloys. *Puṣpāñjan* mentioned in *Caraka-saṁhitā* as well as in *Suśruta-saṁhitā* and used in treatment of eyes and open wounds, is probably zinc oxide. The writings of Manu, Yājñavalkya and Pātañjali of the pre-Christian era sometimes refer to bronze (*kāṁsya*) and brass (*rītikā*) almost synonymously.

During the first century AD, the Roman traders wrote about Indian brass. While the Periplus used the Greek term oreichalcos (mountain copper) for brass, Pliny mentioned the Latin term *aurichalcum* (golden copper), made in India from *cadmia*, identified as calamine or the zinc ore. Samuel Beal suggested that the name *cadmia* came from Calamina, a port at the mouth of the Indus, which negotiated the export of the ore or the alloy of zinc. Ball,[41] however, suggested, that the port was Calliana or Kalyāṇ near Bombay. The sixth-century AD traveller Sopater had mentioned Calliana exporting brass.

Philostratus of Lemons, writing around AD 230, mentioned a shrine in Taxila (presumably Buddhist) in which 'the various figures were portrayed in a mosaic of orichalcum (brass), silver, gold and oxidised copper'. A much more significant fact related to Indian brass of this era is the advent of the alchemist Nāgārjuna.

The earliest literary reference to zinc metal — Nāgārjuna's date

The earliest reference to zinc as a metal is found in Nāgārjuna's *Rasa-Ratnākara* (*RR*). In one passage (*RR*, 3) it was mentiond:

> What wonder is that *rasaska* (zinc or zinc ore) roasted with three parts of *śulva* (copper) converts the latter into gold.

Actually, this was gold-coloured 25 per cent zinc brass, also known as *pīta-tāla* (*pītala*) or yellow alloy. The description given above could have corresponded to mixing of metals or the cementation process. Another passage (*RR*, 31-2), however, leaves us in no doubt that zinc metal was obtained as a reduction product of *rasaka* when roasted with a reducing agent and borax in a covered crucible:

> *mūkamūṣāgatam dhmātam ṭaṅkaṇena samanvitam satvam kuṭilasaṁkāśam patate nātra saṁśayaḥ* ।

> there is no doubt that this (process) yields an essence (metallic) of the appearance of tin.

Al-Birunī of the tenth century AD wrote that "Nāgārjuna lived nearly a hundred years before our time". On this basis, Ray put Nāgārjuna in the eighth/seventh century AD.[42] Filliozat[43] has, however, pointed out that Hiuen Tsiang attested in AD 630 that the alchemist Nāgārjuna had lived much before his time. When his contemporary *Sadvaha-raja* fell short of gold, 'Nāgārjuna Bodhisattva, by moistening all the great stones with a divine and superior decoction, changed them into gold; then *rājā* used this to make four golden statues of Buddha'.[44] *Rasa-Ratnākara* itself mentions *Śālivāhana*. Evidently, what Nāgārjuna did was to suggest the preparation and use of gold-like brass.

Nāgārjuna, the alchemist, was born in Karhād (17°7' N and 74°11'E) of Sātārā district in Maharashtra (parents: Trivikrama and Savitā).[45] Al-birunī put him as a native of Daihak near Somnāth. Nāgārjuna was a contemporary and friend of Yajñaśrī Gautamīputra Sātavāhana or Sātakarṇi (AD 174-203). His friendship with the powerful Sātavāhana king, who had ousted the Scythians from the central and western India, is attested by his famous work *Suhṛllekha* ('Letters to a Friend').

In another text, *Kakṣapuṭatantram,* Nāgārjuna[46] mentioned the levitating pressure of mercury vapour and an alloy *trilauha* of three metals, gold, silver and copper (5:8:6). Presumably, Nāgārjuna's alchemistry, which later influenced many other workers in the sub-continent from Kerala in the south to northern Bengal in the east,[45] was directed towards the substitution of gold. Nāgārjuna might have been aware of the Indo-Pārthian tradition on brass in the second-century AD Gujarat (*vide* item 6 in Table 5.1) and Yavaneśvara's Greek book (AD 150) on mineralogy.[47] The secret art of zinc distillation, practised in Rajasthan and Gujarat, and utilised by Kuṣāṇas, Pārthians and Scythians in brass technology, was passed on to the Sātavāhana civilisation by the knowledgeable alchemist Nāgārjuna. Although Ray thought that the compilation of *Rasa-Ratnākara* in the present form took place in the seventh century AD,[42] the events narrated and the personality of Nāgārjuna clearly belonged to the latter half of second century AD (furthermore, the alchemist Nāgārjuna is not to be confused[48] with the equally famous Mādhyamika philosopher Nāgārjuna, who was probably a contemporary of king Kaniṣka, AD 78-102).

Thus, it may be submitted that the proofs of the earliest (in the world) discovery and use of zinc as a metal in India are of two types: (a) use of brass containing more than 28 per cent zinc in Taxila (reference 32 and data in Table 5.1), and (b) literary reference of Nāgārjuna belonging to the second half of second century AD.

Nāgārjuna boosted the tradition of brass which had originated in India as early as the second millennium before Christ. After Nāgārjuna, we find an extensive proliferation of brass-trade in central India,[49] Gujarat and the rest of the sub-continent. An inscribed brass statue of Buddha, 30 cm high and 13.5 cm wide, of the sixth century AD, has been discovered in a *dharmaśālā* at Fatehpur, a village 20 miles due west of Kāngrā-Koṭ.

The early sixth-century AD text *Bṛhat-saṁhitā* of Varāhamihira (*BS* 57.1-7) described several *vajra-lepa* or cement recipes, one of them containing *rītikā* or calx of brass. During the same century, Sopater, a traveller, mentioned the port of Kalyāṇa exporting brass.

When Hiuen-Tsang visited India (AD 629-645), he found many statues made of *teou-shih* or brass: standing Buddha at Bāmiyan, cast in small pieces and then joined together, the image of sun-god in Multān, statues in the kingdom of Takka on the river Vipāśā, some near the Kulu province, Māyāpura or Haridvāra, Garhwāl and Kumāun areas, brass statues of Śiva in Vārāṇasī, of Buddha in Bodh-Gayā, 'ornamented with rare jewels', and lastly the brass *vihāra* of Nālandā: 'being built by Śilādityarāja (Harṣavardhana AD 606-647) not yet finished, when finished would measure 100 feet'.

Thus, it is clear that by the seventh century AD, India had well-established pre-Muslim traditions in brass-making.

Pre-Muslim era tradition at Zāwar

One small slag heap in the Zāwar area is dated to the seventh century AD,[50] and this is contemporary with the date of the earliest historical references to Zāwar. This reference is on an inscription found at Sanoli, well to the south of Zāwar, which refers to the arrival of a merchant who had a mine at Zāwar, then known as Aranoyagire or hill of wells (indicating earlier mine-shafts in the hilly area).[17,50] There are similar inscriptions on old temples constructed in and around old Zāwar during this period. (Vide Plates 20 & 21)

While Gujarat and other parts of India were making brass icons out of distilled zinc during the period fifth-eleventh centures AD continuously (*vide* reference 30, also Table 5.1), the Muslim world made zinc oxide but not the metal itself. The Iranian Abu Dalaf, writing in the tenth century, describing zinc oxide and its use in brass-making (evidently by the cementation route), stated that in India "it is made from the vapour of tin".[51] Al-birunī of the the eleventh century and the Italian traveller Marco Polo of the thirteenth-century described the Iranian process of obtaining zinc oxide (sublimate deposited on pegs or bars) but not the metal itself.

The earliest evidence for zinc smelting in medieval Zāwar is the C-14 date of 840 ± 130 BP (around AD 1150) for one of the heaps of white ash removed from the zinc-smelting Koṣṭhī furnaces.[50] The main expansion of the industrial phase of zinc production was executed by Rāṇā Lākhā of Mewār (1382-97). Even after the twelfth century AD, the local rulers at Udaipur retained the control of zinc and brass technology at Zāwar. Rāṇā Pratāp Singh (1572-97)) was forced to hide from the armies of Akbar in one of the spacious mines on Zāwar Mālā, now known as

Pratāp Khān (Vide Plate 23). In *Āin-ī-Akbarī* completed in 1596, Abul Fazl admitted his ignorance about *jast* (derived from Sanskrit *jaśada* or zinc): "according to some this is *ruh-i-tūtiyā* * (the metal of *tūtiyā* or zinc oxide): it is nowhere mentioned in the philosophical books (Persian and other Islamic works); there is mine of it in Zāwar".[52]

Discovery of the medieval zinc smelting outfit

It is amazing that nearly seven centuries long (AD 1150-1810) medieval or pre-modern tradition of zinc smelting at Zāwar was nearly forgotten. Evidently, much secrecy was maintained there. Only the iatro-chemist authors of the *Rasaśāstra* texts, such as the thirteenth-century *Rasa-Ratna-Samuccaya*,[3] had some access to the details and described a part of the same.

Colonel Tod, the great historian of Rajasthan, wrote[53] in 1829 that as recently as 1760, the Zāwar mines had been in full production. The mines were abandoned during the reign of Mahārāṇa Bhīm Singh (1778-1828), when the State of Mewār was repeatedly plundered by the Marāṭhā invaders. In 1850, Brooke[54] mentioned the ancient art of zinc distillation and an old man who claimed that he knew the technology.

In the neighbouring villages, the used brinjal-shaped retorts have been utilised as components of building-walls and this has always been a familiar sight, although people forgot how and where these retorts were used. (Vide Plate 22) Attention has been drawn to the ancient deep mine shafts[55] and the huge quantity of zinc extraction debris at Zāwar.[56] (Vide Plate 24)

Straczek *et al.*[56] pointed out that there are two distinct types of smelter residues. One is a sinter or clinker intermixed with fragments of fired clay pots or retorts and contained in unbroken retorts; the retort residue and wall contain substantial amounts of zinc, showing that these retorts are possibly zinc

---

* The referee commented that *tūtiyā* is copper sulphate and not zinc oxide. The word *tūtiyā* is actually a generic name like vitriol, which can be blue (copper sulphate), green (iron sulphate) or white (zinc sulphate). In India, *tūtiyā* commonly represented copper sulphate. Colonel Yule and V. Ball, however, noted that *tūtiyā* was the ancient Persian word for zinc oxide. Therefore, the Persian *ruh-ī-tūtiyā* or essence of *tūtiyā*, as mentioned by Abul Fazl, has been rightly translated by Blochmann as the metal from zinc oxide or zinc itself.

smelting crucibles. The other kind of residue is normal glass slag left after the recovery of lead and possibly silver.

In April 1980, the Hindustan Zinc Limited (HZL) sponsored a three-year research project on recovery of zinc from the ancient slags which was successfully conducted at the Indian Institute of Technology (IIT), Kanpur by the present author.[57] In 1982, HZL collaborated with British Museum Research Laboratory (P.T. Craddock, Lynn Willies, etc.) and the Department of Archaeology, M.S. University of Baroda (K.T.M. Hegde) on archaeological investigations, and this led to the spectacular discovery of the zinc distillation outfit, including furnaces and retorts showing the production strategy. Craddock described the exciting discovery:

> On the third day of the excavation (December 1983), one of the Baroda team (Hegde) spotted the corner of a refractory plate sticking out from a heap of spent retorts beside a goat track in a valley on Zāwar Mālā. . . . With mounting excitement we cleared a small area above and around it to reveal first, the edges of furnace walls, and then the tops of retorts still *in situ*.[10] (Vide Plates 25-7)

Extensive archaeological and archaeo-material investigations followed;[1-26] we now present the summary of the results of the historical experiments performed in two continents.

Ancient retorts and furnaces at Zāwar

During the 1983 excavations, two groups of furnaces were uncovered. A single bank of seven furnaces upon Zāwar Mālā contained small retorts 20 cm long and 10 cm in diameter. In old Zāwar, there was a more extensive arrangement of furnaces using larger retorts (30-5 cm long and 10-15 cm diameter). In both groups, 36 retorts in a 6 × 6 arrangement were contained within the truncated pyramid of each furnace. Thus, no less than 252 retorts were fired simultaneously in a single bank. The retorts were supported vertically on perforated bricks through which the condenser tubes passed into the zinc collectors beneath (*figs*. 5.1-5.3).

The furnaces are in two parts consisting of a zinc vapour condensation chamber at the bottom and a furnace chamber at the top. The two chambers are separated by a perforated terracotta plate. The condensation chamber measures

65 × 65 cm and 20 cm in height. The perforated terracotta plate that separates the two chambers is a composite unit made up of four equal segments of 35 cm^2. It is 4 cm thick, well-baked, and sturdy. Its perforations include circular holes of two sizes: larger ones of 4 cm diameter, each of which is surrounded by a number of smaller holes of 2.5 cm diameter. Within the furnace, the composite terracotta plate was found to be supported on a ledge in the furnace walls on all four sides and a single solid terracotta pillar placed below the junction of its four segments.[16, 22]

Up above the perforated terracotta plate is the furnace chamber, in which 36 charged retorts (*figs.* 5.1-5.2) were arranged, inverted vertically; it may be presumed that 36 vessels were placed, one underneath each retort, to collect the condensed zinc vapour. This arrangement of downward distillation retort with the condensing unit underneath or *distillation per descensum*, is precisely what had been described in *Rasa-Ratna-Samuccaya* text (II. 157-66; IX.48-50). The brinjal-like retorts in Zāwar (*fig.* 5. 3) are also similar to the *vr̥ntākamūṣā* described in *Rasa-Ratna-Samuccaya* (X.22-3). (Vide Plates 29 & 39).

It appears that a cylindrical reed of 1.5 cm diameter was inserted into the retort after it was charged and the funnel part was luted on it (*fig.* 5.3); this is evident from the central hole which is preserved in many retort residues.[13] The reed helped to keep the charge within the retort when it was inverted and placed in the furnace. When the furnace was fired, the reed burnt away leaving behind a cylindrical flow channel for the zinc vapour to flow freely out of the retort.

An outline of the medieval process

In the Zāwar mines, the host rock is dolomitic and arkosic (quartz and felspar) containing a variety of sulphide minerals, such as sphalerite (ZnS) (the principal zinc-bearing mineral), galena (PbS), sulphides of iron, copper, silver, etc.[56]

In Table 5.2 are presented analytical data on Zāwar ore, tail, concentrate from the modern plant and the ancient retort residue. Craddock *et al.*[15,24] suggested on the basis of 'similar Fe/Pb/Ag ratios in the retort residues and in the modern concentrates' that substantial upgradation or beneficiation was done by the ancient smelters by hand-picking or winnowing, so that the retort charge assayed more than 30 per cent in zinc. Then to explain the high proportion of

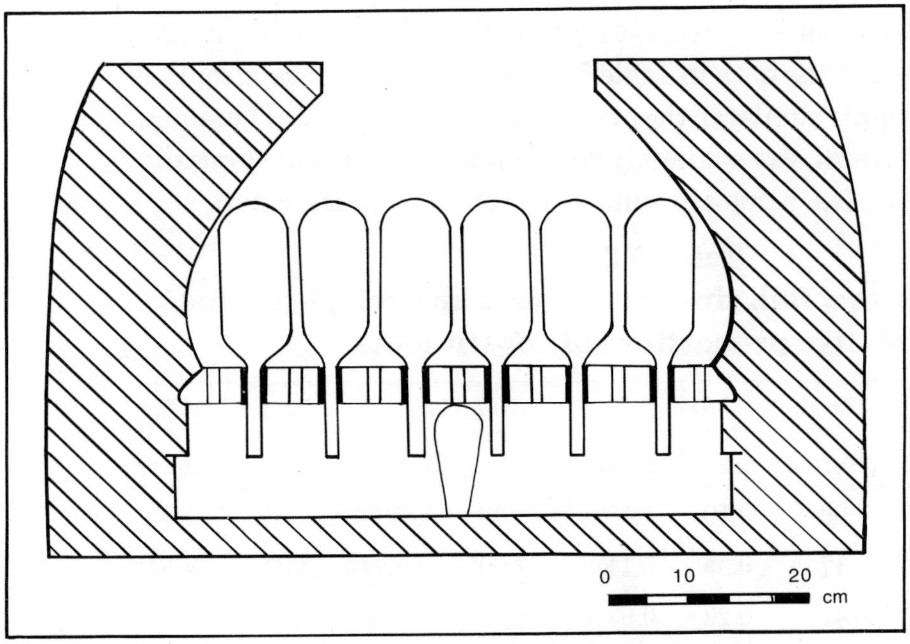

Fig. 5.1: Midpoint vertical section through a zinc reduction-distillation furnace at Zawar Mala (Craddock *et. al.*[13])

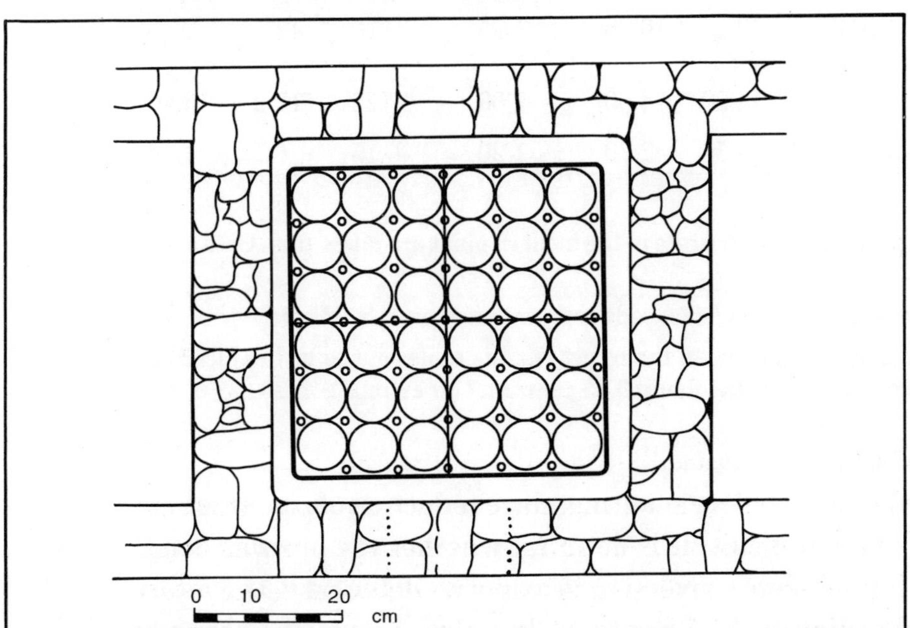

Fig. 5.2: Horizontal section through the furnace in *Fig.* 5.1 at perforated plate level (Craddock *et. al.*[13])

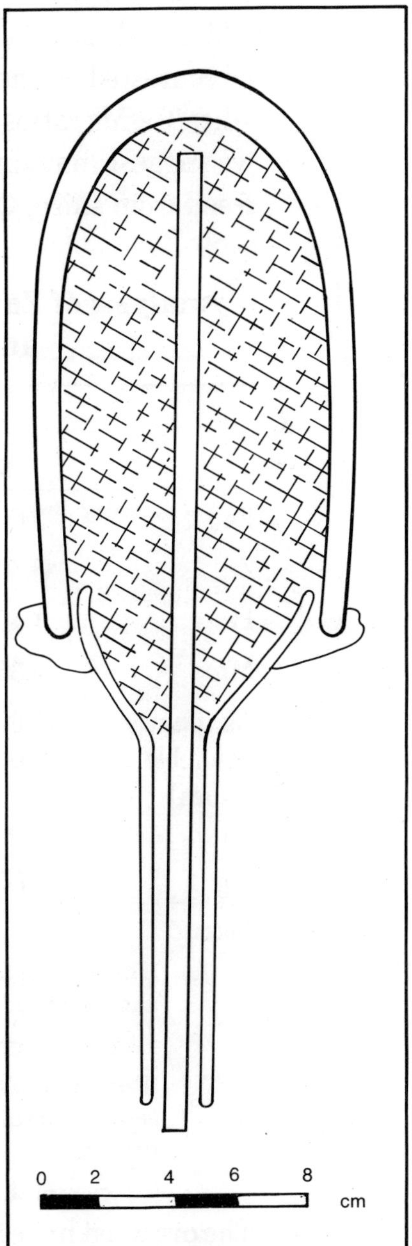

Fig. 5.3: Section through small charged zinc reduction — distillation retort with stick still in place (Craddock *et. al.*[13])

calcium and magnesium in the retort residue, Craddock *et al.*[15, 24] suggested that additional materials like dolomite were added in the charge. This hypothesis may be contested on the basis of the data in Table 5.2 compiled by us. The insoluble silica/Ca/Mg ratios are quite similar in the ore, tail and retort residue. Therefore, we suggest that zinc mineral was not highly beneficiated and not substantially freed from silica, Ca, Mg, Na, K moiety while being charged in the retort.

Table 5.2
Analysis of Zāwar ore, tail, zinc concentrate and retort residue, and relative proportions of constituents

	Mochia Ore[1]		*Mochia Tail*[1]		*Zinc Conc.*[1]		*Retort Residue*	
	%	*Relative*	*%*	*Relative*	*%*	*Relative*	*%*	*Relative*
Fe	3.00	1.00	3.10	1.00	8.00	1.00	5.7	1.00
Zn	3.50	1.17	0.35	0.11	52.00	6.50	3.04[3]	0.53
Pb	1.40	0.47	0.22	0.07	1.30	0.16	1.55	0.27
Mn	0.35	0.12	—	—	0.035	4.4×10^{-3}	1.03	0.18
Ag (ppm)	40.00	10^{-3}	—	—	140.00	1.7×10^{-3}	114[4]	2×10^{-3}
Insoluble (silica)	30.00	10.00	33.20	10.70	1.25	0.16	42.5	7.45
Ca	13.00	4.33	14.00	4.50	1.00	0.125	17.2	3.00
Mg	7.5	2.50	8.50	2.70	1.20	0.15	6.7	1.18

Notes:

1. The data for the ore, tail and concentrate in the modern plant are taken from Chatterjee and Agarwal.[58]
2. Elemental percentages calculated from oxide compositions given by Hegde.[22]
3. From Hegde's data, elemental zinc in the retort residue comes out to be low (1.82 per cent), whereas Freestone[15] gives the figure 3.85 per cent. Our estimate (Biswas)[1] is 3.04 per cent.
4. Average of the values for silver quoted.

The ore must have been roasted before smelting, since reduction of zinc sulphide to the metallic state is very difficult. It is doubtful whether the ore was dead-roasted above 1000°C for complete conversion to oxide; we detected in the retort residue not only a little sulphur (0.35 per cent), but also the specific phase of goslarite or zinc sulphate.[1]

The smelting charge must have included a small quantity of common salt (as surmised from the chlorine and sodium contents in the retort residue) and an adequately large quantity of carbonaceous matter, apart from the calcined ore, and then rolled into pellets of 1 cm^3 volume (*Rasa-Ratna-Samuccaya* or *RRS*, II.163-4 refers to the *guṭikākṛti* pellets containing sodium bicarbonate and borax). The charge (about 1.5 kg per retort) was loaded into clay retorts fitted with funnel-like condenser tubes, as described before (*fig.* 5.3). These were indeed the brinjal-shaped crucible or *vṛntākamūṣā*, as described in *RRS* (II.157, II.163, X.23-4, etc.). On heating in the furnace, zinc oxide was reduced by the carbonaceous matter to zinc vapour. The reducing blue flame of carbon monoxide was observed to be substituted by white flame of zinc vapour, indicating that reduction had taken place (*bhavet nīlā sitā yadi* — *RRS*, II.159-60).

Using a scanning electron microscope and observing the vitrification textures of the Zāwar retort and clay materials, Freestone and Middleton[18] estimated that the temperature reached in the Zāwar zinc distillation furnace was of the order of 1150-1200°C, and that this temperature was maintained for 5 hours or more. The highly endothermic reduction of zinc oxide must have been achieved at a very low partial pressure of oxygen (less than 10^{-20} atm) to prevent re-oxidation of the metal. Zinc vapour condensed in the tube, the temperature being around 500°C, and collected in the vessles placed below. This kind of downward distillation or *tiryakpātana* of zinc vapour, produced under a highly reducing atmosphere, has been described in *RRS* (II.163-8, X.48-50).

A part of the zinc oxide was converted to well-identified silicate phases and thus could not be recovered as reduced metal.

Fresstone *et al.*[14] estimated that 200-500 g zinc was extracted per retort, or 7-18 kg per smelt of 36 retorts. Each retort weighs about 3 kg. Thus, the debris of around 6 lakh tons of spent retorts corresponds to about 1 lakh ton of zinc, according to Freestone *et al.*,[14] which might have been produced at Zāwar during thirteenth-eighteenth centuries AD. This has been indeed one of the most outstanding levels of industrial production in the medieval world.

Phase studies on Zāwar slags at IIT, Kanpur

Six lakh tons of spent retorts contain two lakh tons of residues within, assaying about 3 per cent zinc. Therefore, some 6000 tons of zinc metal remain within the retort residue, and probably another 1000 tons in the lead slag. Our research at the Indian Institute of Technology (IIT), Kanpur, sponsored by the Hindustan Zinc Limited, was directed towards the recovery of zinc from these two kinds of slags.[57] The first step in our work was characterisation of these residues, which turned out to be very useful and relevant to the archaeo-metallurgical problem.

Table 5.3
Phases identified in Zāwar residue (content), wall and (lead) slag by x-ray and electron diffraction studies (Biswas *et al.*[1])

Sample	*Phases identified*	
	X-ray diffraction	*Electron diffration*
Retort content	$Zn_2P_2O_7$, $5H_2O$, goslarite $ZnSO_4$, $7H_2O$, hemimorphite, $Zn_4(OH)_2Si_2O_7$, H_2O, esperite $Ca_2Pb\ (ZnSiO_4)_4$, sphalerite ZnS, Chalcopyrite $CuFeS_2$, larsenite, $PbZnSiO_4$	$Zn_2P_2O_7$, $5H_2O$, goslarite, hemimorphite, esperite, quartz, $Mg_2\ Al_4\ Si_5O_{18}$, hardistonite $Ca_2ZnSi_2O_7$, hydrozincite or aurichalcite, $Zn_5(CO_3)_2)(OH)_6$
Retort wall	Goslarite, hardistonite, hemimorphite, $Zn_2P_2O_7$, $5H_2O$, aurichalcite, quartz, sphalerite	Goslarite, hemimorphite, willemite Zn_2SiO_4, hardistonite
Slag	Quartz, goslarite, $Zn_2P_2O_7$, $5H_2O$, hemimorphite, aurichalcite, sphaleritte, chalcopyrite, hardistonite	Hemimorphite, goslarite, hydrozincite, quartz, sphalerite

The phases identified by X-ray and electron diffraction studies by Biswas *et al.*[1] are summarised in Table 5.3. The results show that in the roasting operation prior to retort distillation, a small part of sphalerite was not converted into oxide and remained in the retort as ZnS and $ZnSO_4$. The presence of goslarite or $ZnSO_4$, $7H_2O$ (hydrated at a later stage) was confirmed by the endothermic DTA peaks apart from X-ray and electron diffraction studies. Some ZnO might have remained unconverted in the retort to undergo atmospheric conversion to basic carbonate at a later stage. While most of the ZnO was reduced to metal, a part of it must have been converted at a high temperature to phosphate and silicates from which the metal could not be recovered. Biswas *et al.* detected a number of zinc, calcium-zinc, lead-zinc, calcium-lead-zinc and magnesium-aluminium silicates.[1] Later, Freestone *et al.*[15] reported that lead slag contained iron, calcium-iron, calcium-zinc, magnesium and calcium-magnesium silicates. The only silicate phase that Freestone *et al.*[15] could report in the retort residue was $CaMgSi_2O_6$ or diopside.

Biswas *et al.*[1] obtained secondary electron images of retort residue samples and found that the particles have a variety of morphology: plates, needles and spheres. Scanning electron microscopy and X-ray microanalysis showed characteristic peaks of many elements, the most prominent being those of silicon and calcium. Scanning was done by a fine electron probe on the zinc-containing particles in the size range 0.5-40 μm.

Several of the small particles (0.5-8.0 μm) show approximately constant values for the ratio of intensities corresponding to the elements Si, Ca, Mg Fe, Al, Zn, Pb, Mn, Na, K and S (in decreasing order of occurrence), indicating that these particles containing a few of the above elements are homogeneous in nature. The larger particles show wide variation in intensity ratios and hence variation in the chemical composition of the grains from point to point. Thus, the approximate size of the zinc-containing and other homogeneous grains in the retort residue is in the range 0.5-8.0 μm.

The size of the particles in the lead slag sample was found to be lower than that in the retort residue. The size of the zinc-containing grains was estimated to be in the range 0.5-6.0 μm, by carrying out point to point analysis. The X-ray spectrum showed the presence of titanium apart from the other elements noted in the retort residue.

The work at IIT, Kanpur was directed primarily towards the recovery of zinc from the siliceous retort residue and slag at Zāwar. We found (Biswas and Kumar[2]) that the non-silicate phase, such as hydrozincite $Zn_5(CO_3)_2(OH)_6$, could be easily leached by acid. A special 'fast leaching' technique, making use of the water-starved nature of the silicate-sulphuric acid system, could recover more than 80 per cent of the zinc contained in the silicate phases, such as hemimorphite $Zn_4(OH)_2\ Si_2O_7$, H_2O, willemite $Zn_2\ SiO_4$, etc.

The characterisation work done at IIT, Kanpur[1,57] has established the existence of complex zinc-phosphate-silicate phases in the slag which could have been produced only by the above 1000°C pyrometallurgical smelting process. This corroborates the conclusions reached by Craddock *et al.*[6-26]

Table 5.4
Brass icons in India (AD 1350-1752)

Sl No	*Site*	*Item*	*Elemental percentages*					*Impurities*
			Cu	*Sn*	*Zn*	*Pb*	*Fe*	
	Medieval Period							
1.	Gujarat AD 1350	Ambikā	68.4	1.6	18.5	9.5	-	Fe, Ag, Bi
2.	" AD 1480	Model Temple with four doors 10 × 24.5 cm	68.6	0.2	28.9	1.6	-	Fe, Mg, Bi
3.	" AD 1485	Viṣṇu-Nārāyaṇa	58.9	0.5	36.8	2.1	-	Fe, Al, Ag, Si, Mg, Bi
4.	Rajasthan 15th-16th centuries AD	Rājpūt Prince on Horse	72.9	0.4	21.8	2.7	1.0	Al 0.3, Ag, Mg
5.	Gujarat AD 1554	Kāla-Bhairava	76.7	1.2	13.8	6.3	1.5	Al, Mg
6.	" 17th century AD	Chauri-Bearer	58.3	1.5	35.5	2.3	-	Fe, Ni, Al, Ag, Cd, Si, Mg
7.	Rajasthan 17th century AD	Dīpalakṣmī Rājpūt girl	58.8	0.9	33.2	4.5	0.8	Al 1.6, Mg
8.	Gujarat 18th century AD	Dīpalakṣmī	52.8	1.1	39.9	2.9	-	Fe, Al
9.	" 1752 AD	Tīrthaṁkara-Seated	62.3	-	36.0	0.5	-	Fe, Ag, Bi

10.	Nepal	Ṣaḍākṣarī Lokeśvara Form of Avalokiteśvara	60.5	2.75	35.3	2.37	-	Fe, Ni, As, Au

Scenario — fourteenth century onwards

The large-scale manufacture and widespread use of zinc and brass in medieval India need to be fully chronicled. Table 5.4 records a few brass icons of India for the period AD 1350-1752; the high zinc content, sometimes in the range 35-40 per cent, in these icons is particularly noteworthy. Item No. 9 in Table 5.4 is dated AD 1752 — five years before the War of Plassey, and eight years before the Zāwar production was slowed down on account of the Marāṭhā invasions. During the medieval period, the Moghuls had used brass (as well as bronze) for manufacturing guns,[35] and artisans of Bidār (83 km from Hyderābād) used high zinc (84 per cent) brass or *bidri* alloy for ornamentation over it by gold or silver ware.[59]

The etymology of words denoting zinc became clearer. *Madanapāla-Nirghaṇṭu* of AD 1374 mentioned *yaśadam vaṅgasadr̥śam* or the 'zinc metal (*yaśada*) like tin' (Table 5.5). *Yaśada* means that which gives *yaśa* or fame; the connection was clear in so far as zinc was known to produce the famous gold-like yellow alloy of brass (*vide* second-century AD text *Rasa-Ratnākara*, 1.3). The European word 'zinc' was probably derived from *yaśada*; the Sanskrit word became *jast* (Abul Fazl) and *dastā* in several Indian languages.

In 1597, Libavius (AD 1545-1616) received Indian zinc, which he called 'Indian or Malabar lead'. He was uncertain what it was. Although Paracelsus (AD 1616) is generally credited to have given the name 'zinc' to the metal, large scale export of the metal from India to the West started later in the seventeenth century, and according to Roscoe, the identification of zinc as the metal from blende or calamine was accomplished by Homberg[35] in 1695.

Table 5.5

Some literature on Zinc (AD 1374-1886)

AD 1374	*Madanapāla-Nighaṇṭu* refers to *yaśadam vaṅgasadr̥śam* (zinc: tin like): *yaśada* means that which gives *yaśa* or fame, converts copper into yellow gold-like brass.

AD 1597	Libavius receives a sample and calls it Indian or Malabar lead.
AD 1616	Paracelsus calls it *zinc* from *yaśada*, in several Indian languages, *dastā*.
AD 1695	Homberg identifies Indian zinc as the same metal from European calamine.
before AD 1730	An Englishman transmitted Zāwar technology to the West — identity not known.
AD 1730	William Champion's experiment at Warmley near Bristol; patent in 1738. Cost of metal £ 260 a ton, whereas calamine cost only £ 6 a ton.
AD 1743	Champion starts manufacturing zinc by distillation per descensum: process 'notoriously close to the Zāwar process' (Morgan and Craddock).
AD 1751	Postlewayt's *Dictionary of Trade and Commerce* admits ignorance about zinc technology. India continues making high-Zn brass statues.
AD 1800-20	Zāwar zinc industry devastated by famine and Marāṭhā invasion.
AD 1886	V. Ball quotes Beckmann's *History of Inventions* (Bohn's Edition, ii, p. 32): "An Englishman went to India in the seventeenth-century to discover the process used there in the manufacture of zinc, and returned with an account of distillation per descensum. I have not yet been able to identity this Englishman".

As late as 1751, Postlewayt's *Universal Dictionary of Trade and Commerce* had to admit ignorance of how the zinc was made in the East. In 1731, imported zinc commanded a price of £ 260 a ton on the London market, whereas the abundant calamine, used to make cementation brass, cost only £ 6 a ton.[13]

Technology transfer to the West

In 1730, William Champion of Warmley (near Bristol, England), started experiments on making zinc metal from calamine by the reduction-distillation route. In 1738, he applied for a patent and in 1743 he commenced production of zinc. Morgan, a modern expert of Imperial Smelting Processes (ISP) Ltd., wondered 'how such a remarkable achievement was completed in so short a time'.[9] Morgan and Craddock *et al.*[13] noted that :

> Champion was notoriously close with details to Indian process at Zāwar; possibly a third party described the general principle of the process to Champion.

The process at Warmley used the same arrangement of retorts and the same technique of 'distillation *per descensum*' as in Zāwar and even included 1.5 weight per cent common salt in the zinc-smelting charge. The closeness was indeed 'notorious'! Craddock *et al.*[13] possibly overlooked the significant remark made earlier on the subject by Ball.[60] Mentioning Beckmann's *History of Inventions* (Bohn's Edition ii, p. 32), and his comments about the earliest efforts for making zinc in Europe, V. Ball wrote:

> Beckmann adds that an Englishman went to India in the seventeenth century to discover the process used there in the manufacture (of zinc), and returned with an account that it was obtained by distillation *per descensum*. I have not been able to identify this Englishman.[60]

Recently, Porter[61] has written : "A Dr. Lane seems to have smelted zinc ore at his copper works in Swansea as early as 1720." Had he visited Zāwar?

It is thus established that there was a technology transfer from Zāwar in India to Bristol in England sometime before 1730. The evidence contradicts the usual notion that the vector of science and technology transfer has always been from the West to the East. Indian traditions have frequently benefited the growths of the Greek, Arab and medieval European sciences. On brass and zinc metallurgy, the primacy of India in the ancient and medieval world is now beyond any dispute.

References

1. A.K. Biswas, A Gangopadhyay and T.R. Ramachandran, Phase Studies on Zinc Residues of Ancient Indian Origin, *Trans. Indian Inst. Metals*, *37* (3), 234-41, June 1984.

2. A.K. Biswas, and Rakesh Kumar, Zinc Recovery from Zāwar Ancient Siliceous Slag, *Hydrometallurgy*, *15*, 267-80,1986.

3. A.K. Biswas, *Rasa-Ratna-Samuccaya* and Mineral processing State-of-Art in the Thirteenth-Century AD India, *Indian J. His. Sci.*, *22* (1), 29-46, 1987.

4. A.K. Biswas, S. Biswas and N.A. Chakravorty, Archaeo-material Studies in India and Literary Evidences, in *Historical Archaeology of India*, ed. by A. Ray and S. Mukherjee Books and Books, New Delhi, 1990, pp. 49-66.

5. A.K. Biswas, *Minerals and Metals in Ancient India — A Few Glimpses,* Centenary Lecture, Bengal Engineering College, 10 April 1992, pp. 1-15, discussed in the July 1992, p. 189 editorial of *Journal of Mines, Metals and Fuels*. The seminar lectures delivered at the Jadavpur University. Aligarh Muslim University and I.I.T Kanpur have been widely appreciated.

6. P.T. Craddock, The Composition of the Copper Alloys Used by the Greek, Etruscan and Roman Civilisation — The Origins and Early Use and Brass, *J. Archaeol Sci.*, *5*, 1-16, 1978.

7. P.T. Craddock, A.M. Burnett, and K. Preston, Hellenistic Copper-Base Coinage and the Origins of Brass. In *Scientific Studies in Numismatics,* ed. by W.A. Oddy. Occasional Paper No. 18, British Museum. 1980, pp. 53-64.

8. P.T. Craddock, The Copper Alloys of Tibet and Their Background, in *Aspects of Tibetan Metallurgy,* Occasional Paper No. 15, British Museum, 1981, pp. 1-31.

9. S.W.K. Morgan, William Champion as a Metallurgist, paper presented to Historical Metallurgy Society meeting at Bath, 20 September 1981; *Trans. Inst. Min. Metall. Assoc. Sect.* C, *92* C 192-C 200. December 1983.

10. P.T. Craddock, Goat Path Discovery Unlocks Secrets of Medieval Zinc, *IAMS Newsletter*, No. 6. Institute of Archaeology, London, 1984.

11. P.T. Craddock, L.K. Gurjar and K.T.M. Hegde, Zinc Production in Medieval India, *World Archaeol.*, *15* (2), 211-17, 1983.

12. Lynn Willies, P.T. Craddock, L.K. Gurjar and K.T.M. Hegde, Ancient Lead and Zinc Mining in Rajasthan, India, *World Archaeol.*, *16* (2), 222-33, October 1984.

13. P.T. Craddock, I.C. Freestone, L.K. Gurjar, K.T.M. Hegde and V.H. Sonavane, Early Zinc Production in India, *Min. Mag.,* 45-52, January 1985.

14. I.C. Freestone, P.T. Craddock, K.T.M. Hegde, M.J. Hughes, and H.V. Paliwal, Zinc Production at Zāwar, Rajasthan, In *Furnaces and Smelting Technology in Antiquity*, ed. by P.T. Craddock and M.J. Hughes, Occasional Paper No. 48, British Museum, 1985, 229-44.

15. I.C. Freestone, P.T. Craddock, L.K. Gurjar, K.T.M. Hegde and H.V. Paliwal, Analytical Approaches to the Interpretation of Medieval Zinc-Smelting Debris from Zawar, Rajasthan, *J. Archaeol. Chem.,* 3,1-12, December 1985.

16. K.T.M. Hegde, P.T. Craddock, and V.H. Sonavane, Zinc Distillation in Ancient India, *Proceedings of the 24th International Archaeometry Symposium*, Smithsonian Institution Press, Washington, 1986, 249-58.

17. H.V. Paliwal, L.K. Gurjar and P.T. Craddock, Zinc and Brass in Ancient India, *Can. Min.*

Metall. Bull. 79 (885), 75-9, January 1986.

18. I.C. Freestone and A.P. Middleton, Mineralogical Applications of the Analytical SEM in Archaeology, *Mineral. Mag. 51*, 21-31, March 1987.
19. P.T. Craddock, The Early History of Zinc, *Endeavour, 11* (4), 183-91, 1987.
20. Lynn Willies, Ancient Zinc-Lead-Silver Mining in Rajasthan, Interim Report, *Bull. Peak Dist. Mines Hist. Soc. Ltd., 10* (2), 81-123, Winter 1987.
21. Lynn Willies, Ancient Lead-Zinc-Silver Mines in Rajasthan *Min. Mag.,* 31-5, January 1989.
22. K.T.M. Hegde, Zinc and Brass Production in Ancient India, *Interdiscipl. Sci. Rev., 14* (1), 86-96, 1989.
23. P.T. Craddock, I.C. Freestone, L.K. Gurjar, A.P. Middleton, and L. Willies, The Production of Lead, Silver and Zinc in Early India, *Old World Archaeometallurgy,* ed. by A. Hauptman, E. Pernicka and G.A. Wagner, Der Anschnitt, Beiheft 7, Bochum, pp. 51-69, 1989.
24. *2000 Years of Zinc and Brass*, ed. by P.T. Craddock, British Museum Occasional Paper 50, London, pp. 1-6 and 29-72, 1990.
25. I.C. Freestone, A.P. Middleton, P.T. Craddock, L.K. Gurjar and D.R. Hook, Role of Material Analysis in the Reconstruction of Early Metal Extraction Technology: Zinc and Silver-Lead Smelting at Zāwar, Rajasthan, *Materials Issues in Art and Archaeology,* II, Materials Reasearch Society, pp, 56-83, 1991.
26. K.T.M., Hegde, *An Introduction to Ancient Indian Metallurgy,* Geological Society of India, Bangalore, A special chapter on 'Zinc and Brass Production in Ancient India", pp. 56-83, 1991.
27. O. Werner, *Spektralanalytische und Metallurgische Untersuchungen an Indischen Bronzen,* Leiden, Brill, 1972.
28. K. Haedecke, Gleichgewichtsverhältnisse bei der Messingherstellung nach dem Galmeiverfahren, *Erzmetall, 26*, 229-33, 1973.
29. S.R. Rao, *Lothal — A Harappan Port-Town* (1955-62) in 2 volumes, Archaeological Survey of India, New Delhi, Vol. II, 1985, p. 660. Chemical analyses of the Harappan level samples were provided by B.B. Lal.
30. Swarnakamal, *Metallic Art and Technology of Gujarat*, Museum and Picture Gallery, Baroda, 1978, pp. 36-7, 56-7, 66, 104-5.
31. R.C. Gaur, *Excavations at Atrañjikhera — Early Civilisation of the Upper Ganga Basin,* Aligarh Muslim University and Motilal Banarsidass, Delhi, 1983, p. 497.
32. J. Marshall, *Taxila*, 3 volumes, Cambridge, 1951.
33. M. Farnsworth, C.S. Smith, and J.L. Rodda, Metallographic Examination of a Sample of Metallic Zinc from Ancient Athens, *Hesperia* (Suppl. 8), 126-9, 1949.
34. R.K. Tiwari, and N.K. Kavdia, Ancient Mining activity around Agucha Village, Bhilwara District, Rajasthan, *Man & Envir., 8*, 81-87, 1984.
35. P. Neogi, *Copper in Ancient India*, 1917, reprinted in 1979 by Janaki Prakashan, Patna,

pp. 40-2.

36. *History of Chemistry in Ancient and Medieval India*, ed. by P. Ray, Indian Chemical Society, Calcutta, pp. 94-5, 1956.
37. E. Lo Bue, Statuary Metals in Tibet, in Reference 8, p. 38.
38. O. Werner, Spectro-chemical Analysis of Antique and Modern Bronzes, *Material Prüfung*, 7 (12), 463-70, 1965.
39. B.B. Lal, An Examination of Some Metal Images from Nalanda, *Ancient India,* 12, 53-7, 1956.
40. U. Von. Schroeder, *Indo-Tibetan Bronzes,* Visual Dharma Publications Ltd., Hong Kong, 1981.
41. V. Ball, Zinc Ores, Their Mode of Occurrence, Metallurgy, and History in India, *Scient. Proc. R. Dubl. Soc.* 5, 321-31, 1886-7.
42. Ref. 36, pp. 116-18. We cannot accept this date nor the other suggestion of Ray that 'brass appears to have been introduced in northern India through Chinese trade' (p. 97).
43. J. Filliozat, Al-Birunī and Indian Alchemy, in *Al-Birunī Commemoration Volume*, Iran Society, Calcutta, pp. 101-5, 1951.
44. Samuel Beal, *Buddhist Records of the Western World*, 1884, Reprint, Oriental Books, Delhi, p. 216, 1969.
45. Sylvain Lévi, Un nouveau document sur le bouddhisme de basse époque dans l' Inde, *Bull, Sch. Orient. Afr. Stud., Lond.*, *VI*, 417-29.
46. Bireswar Banerjee, Siddha Nāgārjuna Kakshaputatantram, *Sci. Cult., 21*(4), 195-7, Oct. 1955.
47. David Pingree (tr. and ed.), *The Yavanajataka of Sphujidvaja*, 2 volumes, Harvard University Press, 1978. Sphujidhvaja's Sanskrit text (AD 270) was a translation of Yavanesvara's Greek text, (AD 150).
48. R.C. Majumdar, *The History of Culture of the Indian People — The Age of Imperial Unity*, Volume II, Bharatiya Vidya Bhavan, Bombay, 1980, pp. 388-9. In this portion, the eminent scholar Nalinaksa Dutta writes on the 'confusion' created by an early Indian source, and sustained in Kumarajiva's biography on Nāgārjuna (in Chinese, AD 405) and subsequent texts.
49. A.S. Bisht and R.C. Jain reported fifth-century AD standing Buddhas in East Nimar, Madhya Pradesh, made of brass 'cast in hot molds by the lost wax process', and assaying 26-27 per cent zinc, *Conservation of Cultural Property in India,* 1981-2, pp. 14-15 and 28-30.
50. Ref. 24, p. 49.
51. J.W. Allan, *Persian Metal Technology,* AD 700-1300, Ithaca Press, Oxford, 1979, p. 43, cited by Craddock, *et al.* (ref. 23).
52. Abul Fazl, *The Āin-ī-Akbarī,* Vol. I, tr. by H. Blochmann, New Delhi, third edition, 1977, pp. 41-2.
53. J. Tod, *Annals and Antiquities of Rajasthan,* Smith and Elder, London, 1829; 1950 edn., Routledge and Kegan Paul Limited, 1950, Vol. 1, p. 399.

54. J.C. Brooke, Notes on the Zinc Mines of Zāwar, *J. Asiat. Soc. Bengal, 19*, 212-5, 1850.

55. A. Mookherjee, Geology of the Zāwar Lead-Zinc Mine, Rajasthan, *Econ. Geol., 59*, 657-77, 1964.

56. J.A. Straczek and B. Srikantan, The Geology of Zāwar Zinc-Lead Area, Rajasthan, *Mem. Geol. Surv. of India, 92*, 47, 1967.

57. A.K. Biswas, Annual Reports of Hindustan Zinc Limited, Udaipur: *Studies on Recovery of Zinc from Flotation Tailings and Ancient Slags*, Indian Institute of Technology, Kanpur, 1981-4.

58. B.N. Chatterjee and B.L. Agarwal, How Zāwar Beneficiates Low Grade Lead-Zinc Sulphide Ore, *Proceedings, International Seminar on Lead, Zinc and Cadmium,* New Delhi, Vol. 1, November 1981, pp. B72-B.97.

59. Syed Jafar Mahmud, *Metal Technology in Medieval India,* Daya Publishing House, Delhi, 1988, pp. 115-26. Jafar Mahmud provides some details about the Bidri ware as well as the Zāwar technology.

60. V. Ball, Ref. 41, p. 322.

61. Frank Porter, *Zinc Handbook,* Marcel Dekker Inc., New York, 1991, p. 60.

6

Bidri and Other Art Works in Pre-Modern India

In pre-modern India several traditions of art works based on metals, alloys, gems and stones flourished and became internationally famous. Many of these traditions started in ancient India and continue vigorously in modern India.

Bidri ware, the sleek and smooth dark-coloured metal work with intricate eye-catching designs on its glossy surface, is famous all over the world. This metal work as well as the technique to produce it are found in India alone.[1-5]

Bidri is an alloy which contains 76 to 98 per cent (normally in the neighbourhood of 95 per cent) zinc, 2 to 10 per cent copper, up to 8 per centage lead, 1 to 5 per cent tin and trace of iron (*vide* Table 6.1). Occasionally high per centage (19.9) of lead and (11.4) tin have been noted. However mostly it is high zinc low copper alloy. Up to 1 per cent copper in the zinc forms the terminal solid solution η; above 1 per cent copper the $\in$ phase precipitates at the grain boundaries in this phase field. The usual yellow brass may contain not more than 40-50 per cent zinc, often less; copper constitutes the remainder or the predominant phase. Thus brass and bidri represent the two opposite ends of the zinc-copper phase diagram.

The Bidri ware's surface is first made smooth and a solution of copper sulphate applied to it to darken it temporarily for engraving. The engraving tools cut the intricate but delicate tapestry of design into the metal which is then lighter in colour than the darkened surface and enables the pattern to be seen clearly.

Table 6.1

Results of atomic absorption spectrometry analysis of Bidri ware (taken from La Niece *et al.*[4])

Victoria & Albert Museum acquisition nos.	*Description*	*% by weight content*				
		Zinc	*Copper*	*Lead*	*Tin*	*Iron*
2539-1883 I.S	Huqqa	91.2	3.2	2.9	1.2	0.2
1578-1904	Huqqa	98.0	3.3	1.4	0.1	0.2
45-1905	Huqqa	88.2	3.0	2.9	0.7	0.5
I.S. 46-1977	Weight	92.6	4.7	1.4	5.7	0.1
1479-1904	Ewer	89.3	4.8	0.8	<0.1	0.8
I.S. 10-1973	Bowl	92.1	3.6	1.4	0.3	0.1
02942 (I.S.)	Huqqa	84.6	3.3	1.0	0.1	0.5
I.S. 131-1958	Pan box	95.0	3.1	1.5	0.1	0.1
I.S. 181-1965	Huqqa	76.1	10.1	8.2	11.4	0.6
I.S. 31-1976	Bottle	98.6	2.6	1.0	0.4	1.2
857-1874	Huqqa	92.1	3.6	0.8	0.6	0.4
I.S. 17-1970	Pan box	81.3	2.0	2.0	<0.1	0.5
02949	Bottle	83.8	2.5	6.6	<0.1	0.2
02949	Bottle	93.2	2.4	1.1	0.1	0.2
855-1874	Huqqa	99.2	2.6	0.4	0.2	0.6
02941 (I.S.)	Basin	91.2	2.8	2.1	0.1	0.2
I.S. 4-1977	Huqqa	83.6	3.7	0.8	<0.1	0.6
I.S. 19-1978	Huqqa	85.3	4.6	3.0	1.3	<0.1
120-1886	Bottle	86.7	5.3	1.9	0.2	0.1
I.M. 224-1921	Bottle	91.7	3.6	0.7	0.7	0.7
2066-1883	Box	79.6	2.7	19.9	0.1	0.5
1402-1903	Huqqa	95.3	2.7	1.1	3.0	0.5
I.S. 11-1973	Vessel	97.6	3.4	1.8	<0.1	0.1
I.S. 39-1976	Huqqa	89.2	3.4	0.5	0.7	0.7
I.S. 19-1980	Huqqa	80.6	3.6	7.3	6.4	0.9
856-1874	Huqqa	77.4	4.4	0.9	0.4	0.1

The piece is then handed over to the inlayer. The inlay may be of silver, brass or occasionally gold. The final stage, after the inlay has been burnished, is to blacken the surface of the piece so that the inlay stands out. This is done by applying a paste of ammonium chloride, potassium nitrate, sodium chloride, copper sulphate and mud which darkens the body by producing a characteristic black patina while having no effect on the inlay. The paste is later washed off and finally oil is rubbed into the piece to deepen the blackness of the patina. The result is a lustrous dense black body to contrast with the shining lining — white (silver) or yellow (brass or gold). (Vide Plates 34 & 35)

Replication experiments and chemical hypothesis

La Niece and Martin[4] performed some replication experiments to show that the black colour of the patina was due to copper. The recipe for blackening the Bidri metal has the main constituents in a warm solution of potassium nitrate (one part) often substituted by well-urinated soil, and ammonium chloride (four parts).

In a replication experiment a clean pure zinc sheet was immersed in the above solution, when a *pale grey* patina of zinc oxide and chloride was produced. When the experiment was repeated with the addition of copper sulphate, a reasonably good but *superficial black* patina formed almost instantaneously. Again only zinc oxide and chloride were the main crystalline products. XRD analysis identified $Zn_5\ (OH)_8\ Cl_2$, H_2O, ZnO, Zn and Cu_2O (cuprite) as the crystalline phases. None of these explains the blackness of the patina which was amorphous and contained copper.

Similar experiments were performed with a specially prepared alloy of zinc containing 3 per cent copper. When this alloy was dipped in a warm solution of potassium nitrate and ammonium chloride, it turned as deep black instantly and the patina was even and *adhered well* to the metal surface. This was found to be non-crystalline. Scanning electron microscopy showed the patina to be about 10 μm thick assaying 30 per cent copper as a contrast to 3 per cent in the bulk. Zinc and chlorine were the other elements detected.

La Niece and Martin have postulated[4] that ammonium chloride preferentially dissolves the zinc from Bidri and the resulting copper-enriched surface or the copper-rich ∈ phase precipitate gets oxidised by potassium nitrate producing the black colour. The use of clay does not seem to be crucially important. It could merely serve as a source of alkali nitrate and as a poultice to absorb the unwanted zinc chloride formed during the process. The matt black patina is easily damaged by the standard cleaning techinques designed to remove the white decomposition products from the zinc.[2]

History of the art tradition

The mystery of the black patina is not yet fully solved. The inventors of the tradition did not have any idea about the underlying scientific principles; they merely hit upon the arts and crafts aspect of the process.

The craft of Bidri ware is a kind of damascene work which has been defined by Sir George Birdwood as

> the art of encrusting one metal on another, not in *crustae*, which are soldered or wedged, but in the form of wire, which by undercutting and hammering, is thoroughly incorporated into the metal which it is intended to ornament.

The orginal tradition at Damascus was to encrust gold wire, and sometimes silver wire, on the surface of iron, steel or bronze.

A group of the damascene craftsmen moved from Syria or Iraq to India. Some of them were at Ajmer in Rajasthan and hit upon the idea that damascening could be done on the base of high zinc low copper alloy. Zāwar in Rajasthan was the major zinc-producing centre in the medieval world. The said craftsmen moved down south during the fifteenth century AD and settled at Bidār (17°55' N and 77°32' E) near Hyderābād. When the art flourished in that place for centuries, it became known as Bidri ware crafts.

The earliest known craftsmen like Abdullah-bin-Khaiser and his pupil Sivanna worked at Bījāpur. Historical evidences indicate that the beautiful articles presented to Alāuddin Bahamanī II (AD 1434-57) on the occasion of his coronation impressed him very much, and he invited the craftsmen of Bījāpur to settle at Bidar itself. The workers used to prepare beautiful huqqa base, ewer, bowl, pan box, basin, bottle, slabchis, cot legs, etc. The varieties of workmanship of the design consisted of *tara kashi* (inlay of wire), *taihnishan* (inlay of sheet) *zaranishan* (low relief, inlay levels with surrounding area), *zarabuland* (high relief, for example silver over a lead pad), *āftābī* (design in overlaid metal sheet), etc.

The Russian traveller Althanasins Nikitin, who visited Bidār during AD 1470-74, took with him some of the early Bidri ware specimens for presentation to the Russian Emperor. A large number of articles of Bidri ware were made for presentation to the Prince of Wales when he visited India in 1875. These now adorn the collection of the Victoria and Albert Museum which has published a comprehensive bibliography and illustrated catalogue on the subject.[1] Bidār and Hyderābād museums also have beautiful collections of this kind of ware.

Mahmūd[5] has given some details about this craft under specific headings such as raw materials, tools, implements, process of production, preparation of alloy, mould making, casting (such as a goblet), etc. The manufacture of Bidri ware has been carried on under a system of division of labour. The moulder prepares the alloyed metal, casts the vessel and turns it to its proper shape by his lathe. The carver engraves the patterns on the surface of the vessel, and the inlayer designs the patterns, inlays the ornament of gold, silver or brass, and finally polishes the article. In the pre-modern India there have been four notable seats of Bidri ware manufacture: Bidār, Lucknow, Purneā (Bihar) and Murśidābād (Bengal).

Without steam or machinery

Modern Indian industries started with machineries, and therefore Pramatha Nath Bose rightly defined his treatise on pre-modern 'Art Industries' in India as one dealing with industries 'carried on *without the help of steam or machinery* except of the simplest kind, and which have a remote, if any, connection with natural science'.[6] It is a matter of national pride that such art industries not only survived through the ancient and medieval periods but also remain in prestige in the modern industrial world and earn foreign exchange. We may briefly summarise some of those traditions as they existed during the eighteenth/nineteenth centuries. Since our studies are related to metals, alloys, minerals, gems and stones only, we naturally leave out the non-mineral/non-metal traditions.[7]

Having discussed Bidri, a high-zinc low-copper alloy and its uses, a logical sequence demands that we refer to the use of brass, a medium-zinc high-copper alloy and of copper itself. Fourteenth-century AD *Madanapāla-Nirghaṇṭu* recognised zinc (*yaśada*) as a separate metal and sixteenth-century AD *Bhāvaprakāśa* termed brass as an alloy (*upadhātu*) of zinc and copper.

Brass and copper vessels were normally used in India for domestic purposes instead of porcelain, glass and silver ware. Hindus preferred sacrificial utensils, cooking utensils, plates, cups, drinking vessels and idols made of brass. In Bengal, copper was also preferred for sacrificial vessels; bell-metal or bronze containing copper and tin was not considered pure enough for such purposes. However known as *phūl* in upper India and *kāṁsā* in Bengal, bell-metal was employed in the manufacture of plates, cups and drinking vessels.

During the medieval period, guns were made of wrought iron (forge-welded), brass and bronze (cast). The 14 ft long and 22½ inch bore 1469 maunds weight 'Great Gun of Agra' was a Moghul gun made of brass. The *Mālik-i-Maidān* was a gun cast at Ahmednagar in AD 1548 and described as the 'largest piece of ordnance in the world'. The material cast was bronze containing 19.57 per cent tin. The gun measured 14'3" in length and had the enormous diameter of 4'10" at its mouth. (Vide Plate 31) These gigantic objects signified the unique casting skill of the Indians, although these were not art objects !

Coming back to the topic of art objects, vessels of brass and bell metal were made in eighteenth/nineteenth-century Bengal, in places such as Khānkrā near Murśidābād, Jhānjharpur near Darbhaṅgā, Kañcannagar in Burdwān, Islāmābād in Dacca, Bānsberiā in Hughlī, etc. The Khānkrā vessels were prized for their fine shape and polish. The sacrificial vessels were often decorated with floral designs and figures of divinities.

Patna used to produce a peculiar kind of brass tea-urn. Vārāṇasī brass ware with its excellence of the cast, rich colouring, gold-like lustre, etc., was probably the best in India. The brass was first moulded into the required shape, and then patterns were engraved. Plates, water-goglets, trays, cups, salvers, shields, betel-holders, etc., from Vārāṇasī used to be sold not only throughout India but even all over Europe. Morādābād ware manufacture started a bit late, after 1876, and then became, like the Vārāṇasī ware, universally admired. Jaipur used to produce ordinary household utensils, smoking bowls, called *gargarās*, other articles with arabesque designs from old Indian patterns in *repoussé*. The Jaipur men were known to imitate any thing given to them.[8]

Tin-coating of metallic utensils

Sometime in the pre-Christian era, Yājñavalkya noted that tin and lead vessels could be cleaned by alkali and acid water, whereas copper alloys like *kāṁsya* (bronze) and brass should be cleaned by ash and water (*Ācārādhyāya*, 190). Thus, acid-resistance of tin was known in ancient India. Bodhāyana referred to solid tin vessels. It is not known whether tin-coating of metallic utensils used to be done in ancient India for protecting such utensils, particularly cooking vessels, from chemical action caused by sour or acid food placed in them. During the medieval period however, this became a well-known practice.

Gode has reviewed the literature[9] and opined that this tradition of tin-coating was started in India during the Muslim period. Along with Bahmanī coins (AD 1347-1500), a copper contatiner with both internal and external coating of tin has been discovered in the archaeological excavations at Brahmapurī. Sanskrit texts of the period 1676-1710 show tin as *kalhai* / *kalāya* and tinning as *kalāya-lepa*. The word *kalhai* or *callai* was Arabic, and it might have been derived from Sanskrit *kastīra* or Greek cassiteros.

The Muslims were fond of tinning the cooking vessels — not only the inside portion as Hindus did, but also the outside part. Abul Fazl reported that the copper utensils of Akbar's personal use were tinned twice a month whereas those of the princes and the Harem only once in that time. During the early part of the seventeenth century (AD 1600-50) the British East India Company sold a lot of tin to the Indians, the Arabs and the Persians. With the discovery of new tin ore reserves in Malayasia, the British trade dwindled. Major uses of tin in the Muslim world was in making bronze, bell-metal and tinning vessels of copper and copper alloys.

Tinning was popular in the Hindu medieval world also. The saint-poet Sūrdās wrote: *āi udharī prīti kalaī* comparing love with tin-coating. *Śivatattvaratnākara* (*c.* AD 1710) wrote *kalāye-lepita pātre* (in a tin-coated vessel), etc. *Rājadharmakośa* (*c.* AD 1676) recorded *kathilaṁ vaṅgamucyate . . . kalhaikāraḥ sīsakāro*, etc.

Kathila, *kalāya*, *kastīra*, *vaṅga*, *raṅga*, etc., were different names for tin. During Śivājī's time *kalhāikāra* or *kalhaigār* was probably *sīsakāra* also, i.e., the same people were practising the art of tin-coating as well as lead-coating.

The Bombay Gazeteer of 1904 contained notes on *kalaigar*, a caste of Muslim tin-smiths, in the districts of Ahmadnagar, Pune, Satārā, Śolāpur, Belgaum Dhārwār, Bījāpur and Nāsik. The present author (of this monograph) recalls his childhood days of the late 1930s and early 1940s when *kalaiwālāh*s were frequent visitiors to the households in Calcutta. As Gode observed, they moved about from house to house, doing the work of tin-coating of copper and copper-alloy vessels on the spot with the help of instruments and materials which they carried with them. They had regular shops in the cities also. Their trade dwindled with the increasing use of aluminium and steel for making cooking vessels and other household items.

Gold and silver wares[10]

Indian women have always been fond of jewelleries, the poor amongst them buying items made of lac, glass or brass — and the rich preferring silver and gold ornaments. The finest sorts of Indian filigree have ever been famous. Many of these ornaments were set with precious stones and beautifully enamalled.

Bose reported[10] artisans of Dacca and Cuttack making beautiful gold and silver plates, filigrain jewellery and other filigrain works. Some of the articles used to be made were scent-holders, rose water sprinklers, card-cases, *hukkā*s or smoking bowls, etc. Silver plates made at Bhowānīpur in Calcutta attracted considerable notice. The frost silver and items like tea-sets, saucers, etc., catered to the European taste. In silver wares the ornaments were engraved, while in others they were beaten out (*repoussé*). Kapūrthalā State was famous for vessels of glass covered with silver work. Toṅk in Rajasthan used to produce gold and silver plates 'good in design as well as in execution'. Larger articles, such as thrones and staves of canopies, were made at Jaipur in precious metals. The Mahārājā of Alwar had a number of silversmiths who made tumblers, cups, saucers, tea-pots, *hukkā*s, betel-holders, etc. Bikaner had sheltered merchants and artisans fleeing from the pillage of the later day Moghuls and Mahraṭṭas, and they in turn initiated excellent traditions in silver and gold art objects.

No part of India was however more celebrated for its work in precious metals than Cutch in Gujarat. The Cutch silverware was internationally famous. There was originality in the form of trays, scent-holders, betel-boxes, water-goglets, cups and other articles made by the goldsmiths and silversmiths of Mysore.

Gold and silver wires were used in lace-making, and *kālābatun* (gold or silver wire twisted with silk thread) in the weaving of brocades and cloths of gold and silver. Silk fabrics with raised patterns were called *brocades*. Gold or silver cloths, i.e., silk woven with gold or silver thread, were known as *kinkhabs*. Silk brocades were made wherever silken stuffs were manufactured on an extensive scale, e.g., Murśidābād, Vārāṇasī, Multān, Ahmedābād, Sūrat, Pune, Auraṅgābād, etc. *Sārīs* made at Baluchar near Murśidābād, bedecked with flowers and figures were highly appreciated by the Bengālī ladies. In Bombay and south India, gold wire was introduced as a border in superior fabrics made of cotton or silk. In the *kinkhabs* however gold or silver was worked on a silk base all through the piece 'practically making it in all appearance into a cloth of gold or silver'. Often 'gold' wires used were really silver wire with a gold coating or even gilt copper wire. The gold and silver brocades were made at Murśidābād but the Vārāṇasī brocades were more popular; the varieties woven there were numerous. It was estimated that upwards of 2750 workmen found employment in the manufacture of silk fabrics and silver brocades in Vārāṇasī. Ahmedābād and Sūrat were also famous for the *kinkhabs*. A piece presented to the Prince of Wales by the Gaikwar of Barodā was described 'as one of the most sumptuous ever seen in Europe: this was of Ahmedābād work, rich with gold and gay with colours'.[10]

Purification and recovery of precious metals

While using gold and silver for making jewelleries and other articles, there used to be a lot of wastage. Over the centuries, Indians evolved methods for recovering, purifiying and re-using precious metals. Ṭhakkura Pherū[11] and Abul Fazl[12] had described crude techniques for purifying and assaying gold and silver objects.

The *salonī* process of assay involved repeated treatment of thin plates of impure gold in fire along with saltpetre and brickdust. The final product was pure gold, and the ash (*salonī*) contained all silver, as chloride or nitrate, and some gold. Gold could be recovered by an amalgamation procedure; smelting with lead gave recovery of silver. The use of the process necessitated small imports of mercury and lead into medieval India. The *salonī* process of assaying impure gold was followed for centuries, even by the Calcutta goldsmiths as late as 1827 and the Nepalese goldsmiths as late as 1834 (*JASB*, Decemher 1834, pp. 622-8).

Since the days of Abul Fazl, improvements were made in the techniques of recovery, purification and assay of the precious metals. During the end of the nineteenth century the processes were 'still carried out by more or less illiterate people of the artisan class'. A contemporary status report was provided by Acharyya Prafulla Chandra Ray.[13]

The black colour in impure gold was often due to its copper content present as cupric oxide. A simple step of purification consisted of heating the impure metal in a charcoal fire to redness (when copper oxide was reduced) and then suddenly cooling the metal by sprinkling water so that copper could not get re-oxidised. Copper oxide was also dissolved when impure gold was treated in boiling decoction of unripe tamarind which contained tartaric acid. Zinc oxide in the soldered surface was removed by putting a paste of common salt and potash alum, placing this on fire and then washing with water. Alternatively, the impure metal with white solder was treated in a boiling solution of nitre, common salt and alum. The evolved chlorine and nitrosyl chloride acted upon zinc as well as gold, when some gold was lost in solution as chloride. The fluid containing dissolved gold was known as *jamak* which used to be purchased and treated for recovery of gold by a class of people called *jamakwālā*s.[13]

Following the previously stated nitre-common salt- alum treatment, purified gold metal could be further purified by repeating this treatment, this time with the addition of gradually increasing dose of finely divided sulphur. Chlorine and sulphur chloride produced a little bit of auric chloride which on heating decomposed into aurous chloride and then aurous oxide, which is violet in colour. This rendered a thin violet layer on the surface of the articles.

Having described the pre-modern techniques of purification of impure gold, let us now mention two important pre-modern traditions of *recovery* of gold from the waste products. During the eighteenth/nineteenth century (and may be even at earlier times) there were two categories of people earning their livelihood by recovering gold from the refuse. They were known as *neharwālā* and *jamakwālā*.[13]

The *neharwālā*s collected the daily sweepings from the goldsmiths' shops (*nehar* means sweepings). The heaps of dust and refuse were fractionated by gravity separation yielding heavy metallic particles settling at the bottom. The metallic particles were dried and melted in a large crucible. The metallic button thus produced, used to be purified in the olden days by the *salonī* process described before. Gradually, the *salonī* process was replaced by a step using strong nitric acid. The step was preceded by melting the alloy and then suddenly dropping the molten metal into water when fine powder of the metal was produced and this was subjected to the nitric acid treatment.[13]

*Jemakwālā*s used to collect *jamak* or the waste liquor which was obtained during the purification of gold as well as the cleaning and colouring processes involving gold ornaments. The *jamak* or the waste liquor was found on analysis (by Acharyya Prafulla Chandra Ray) to contain chlorides, nitrates and sulphates of silver, copper, zinc, gold, aluminium, potassium and sodium. Traces of iron and free chlorine were also found. Evaporated residues were found to consist of crystals of alum, nitre, common salt, long transparent needles of $NaAuCl, 2H_2O$ and fine particles of free gold.[13]

The *jamakwālā*s used to boil the waste liquor and the evaporated residue was mixed with a little borax and a large quantity of *poonoor*, a mixture of partially oxidised lead, copper and zinc with traces of silver and iron. *Poonoor* was the by-product obtained during the cupellation-type purification process for silver using a large quantity of lead, as practised during the time of Abul Fazl.[12]

The *jamak* residue, borax and *poonoor* were mixed and made into balls of about 2-3 inches in diameter with the addition of sufficient cow-dung. The balls thus prepared were put into a hearth and powerful charcoal fire aided with large pairs of bellows. The metallic mass produced at the bottom of the hearth had a large percentage of lead. As the blowing was continued, lead was oxidised to litharge and gradually blown off, leaving behind a mixture of gold, silver and copper.[13]

Mīnākarī or the art of enamelling

The art of enamelling or fixing colours by melting in fire has been practised in India ever since the ancient times. During the medieval period, there were many techniques employed in colouring ornaments and other objects of gold and silver, and also other metallic objects made of copper, brass, etc.

In general, mixtures of 10 parts of lead and 3 parts of tin were oxidised by continued heat and exposure to air. To the mixed oxide were added 10 parts of powdered quartz and 10 parts of common salt, and the mixture melted in crucibles. This produced a white enamel consisting of soda-lead glass with tin oxide in suspension, and forming the basis of coloured enamels. Other metallic oxides were added to this for colouring.

The oxides of lead or of antimony produced a yellow enamel. Red enamels were obtained by a mixture of the oxides of gold and iron. The oxides of copper, cobalt and iron gave green, blue and violet respectively, and a variety of intermediate colours by mixtures.[13]

The art of enamelling (*mīnākārī*) jewelleries, etc., was practised by a special class of artisans known as *mīnākāra* or *ruṅgwāllāh*. Jaipur was most famous for enamelling work on gold, followed by Vārāṇasī, Alwar and Delhi. "The colours employed by the Jaipur artists rivalled the tints of the rainbow and there was never a want of harmony." Some Jaipur enamel was of the *champlévé* variety, that is, the outlines were formed from the base itself (plate, etc.), and the colours were deposited in depressions of it.

The enamelling of gold in Multān, Jhang and Kāngrā was generally of a dark and light blue colour, the blue vitreous enamel being the most common. Multān, Bahāwalpur, Kashmir, Kāngrā, Kulu, etc., had traditions of enamelling on silver. Enamelling was done on copper in Kashmir, Jaipur and many other places. Jorhāṭ in Assam produced golden ornaments such as lockets, ear-rings, bracelets, necklaces, etc., enamelled blue, green and white.[6]

Jaipur was famous for enamelling work not only on golden ornaments but also on other wares made of this precious metal. The enamel workers at Jaipur were originally brought from Lahore by Mahārāja Mān Singh. The crutch staff on which he leaned when he stood before the throne of the Emperor Akbar at the close of the sixteenth century has been vividly described:

> It is fifty-two inches in length, and is composed of thirty-three cylinders of gold arranged on a central core of strong copper, the whole being surmounted by a crutch of light-green jade set with gems.
>
> Each of the thirty-two upper cylinders is painted in enamel with figures

of animals, landscapes and flowers. The figures are boldly and carefully drawn, colours wonderfully pure and brilliant and the work executed with skill and evenness.

The round enamelled plate of Jaipur presented to Prince of Wales took four years to complete and was described as 'a monument of Indian enameller's art'.

The artisans of Kashmir enamelled works in silver, copper and brass. The Kāngrā enamel was remarkable for its excellence of blue. The tradition of green enamel made at Pratāpgarh in Rajasthan was a secret jealously kept by two or three families who practised the art.[6]

Encrusted and damascened wares

The Tanjore metal work consisted of soldering, wedging, or screwing on silver patterns, and figures of Hindu deities on copper vessels. The white figures, in the famous Swami style, on red copper ground reproduced an effect at once bold and striking.

Damascening was the art of encrusting one metal on another, not in crustae but in the form of wire hammered into the groove of the base. The original industry at Damascus was limited to encrusting gold wire, and sometimes silver wire, on the surface of iron, steel or bronze. The tradition was brought to India directly from Kabul and Persia. Several areas around Siālkoṭ were the chief seats of this industry. The art also flourished at Lahore, Multān, Jaipur, etc.[6]

The use for which it was originally invented was the decoration of arms and armour, shields and swords, buckles and breastplates, maces and matchlocks. With the disappearance of the ancient art of fighting, the art also got antiquated. We have earlier discussed a special kind of damascened art material namely the Bidri ware.

Another art tradition in India that survived through the ages was setting of precious stones on jewelleries. Skilful *murassiākār* or jewel-setter used to set with unrivalled ingenuity precious stones on massive golden jewellery. The principal stones used were diamond, ruby, onyx, carnelian, emerald, turquoise, jade, agate, jasper, etc. After the goldsmith finished his work, the article went to the enameller and then to the jewel-setter. An Āgrā art-tradition consisted of inlyaing on white marble, ground-coloured stones such as jasper, heliotrope, carnelian, chalcedony, etc.

Concluding remarks

This chapter has been related to the *art* works in pre-modern India based on metals, alloys and gem minerals. It is surprising how Indians have marketted through ages, art objects made 'without steam or machinery', and how some of these traditions are still surviving and likely to be continued through the twenty-first century. However, it must be conceded, that material and metal technology could not enter into the modern age without steam or machinery. Furthermore, the Indians of the eighteenth/nineteenth centuries lacked the elementary knowledge of material *science*, mineralogy and metallurgy which they had to learn from the Western traditions.

References

1. S. Stronge, *Bidri Ware, Inlaid Metal-work from India*, Victoria and Albert Museum, London, 1985.
2. O.P. Agrawal, Bidri, *Museum*, XXVII, 1975, p. 193.
3. T.R. Gairola, Bidri Ware, *Ancient India*, *12*, 1956, pp. 116-18.
4. S. La Niece, and G. Martin, The Technical Examination of Bidri Ware, *Studies in Conservation*, *32*, 1987, pp. 97-101.
5. Sayed Jafar Mahmud, *Metal Technology in Medieval India,* Daya Publishing House, Delhi, 1988. Chapter 14 (pp. 115-26) is devoted to 'Bidri Metallurgy'.
6. Pramatha Nath Bose, A History of Hindu Civilisation During British Rule, First Published, London and Calcutta, 1894, Three Volumes, Reprint, Asian Publication Services, New Delhi, 1975, Vol. II, Book IV, Chapter II on Art Industries, pp. 222-85.
7. Bose has discussed ivory-carving, pottery, glass, leather, cotton, silk, woollen fabrics, dyeing and calicoprinting, embroidery, carpet industries, etc. *Vide* Ref. no. 6, pp. 264-78, 281-3.
8. Ref. No. 6, pp. 254-8.
9. P.K. Gode, History of Tin-Coating of Metallic Utensils in India (AD 1300-1900) *in Studies in Indian Cultural History*, Visveshvaranand Vedic Research Institute, Hoshiarpur, 1961, Vol. III, P.K. Gode *Studies* Vol. VI, Part II, pp. 113-17.
10. Reference No. 6, pp. 238-49, 278-81.
11. Bhamarlal Nahta (ed.), *Dravyaparikkha aur Dhatutpatti* by Ṭhakkura Pherū, Vaishali, 1976.

12. H. Blochmann (tr. and ed.), *Āin-ī-Akbarī* by Abul Fazl, Vol. I, Reprint by Oriental Books Reprint Corporation, New Delhi, 1977.
13. Prafulla Chandra Ray, *History of Hindu Chemistry*, ed. by P. Ray, Indian Chemical Society, Calcutta, 1956, pp. 218-24.

7

Concluding Remarks

MOST Indian traditions related to the extraction and processing of minerals, gems and metals had originated during the 'ancient' period ending in AD 1200, and we had given an exhaustive account of the same elsewhere.[1] The various gemmology texts (*Ratnaśāstra*) and the mineralogy/metallurgy texts (*Rasaśāstra*) described the advances related to pearl, coral, chalcedony, agate, beryl, diamond, iron, steel, copper, lead, zinc, etc.

No doubt some of these texts were compiled during the medieval period. But these were merely the culminations of the earlier efforts. There does not seem to be any evidence that very many ideas diffused from the Muslim intellectual world, from which even the Muslim invaders and rulers in India were cut off. The gap between the Hindu and Muslim traditions was deep, wide and seemingly unbridgeable. For example, Abul Fazl had no concept how the Hindus were making *jast* (*jaśada* or zinc) at Zāwar: 'it is nowhere mentioned in philosophical books (*Persian and other Islamic Works*)'.[2]

Some ideas however did flow into the sub-continent. Thus we find continued and varied uses of gem minerals and adoption of new (Persian/Arabic) names of gems, minerals and metals (chapter 2). The Hindus had considered relative hardness as an important property of minerals to be evaluated. Al-Birunī recommended specific gravity as an additional significant property. Abul Fazl compiled specific gravity data on minerals and metals. Ṭhakkura Pherū, a thirteenth-century AD Jain scholar, was an assay master in Alāuddin's court. We do not know whether he learnt some of the techniques of assaying of precious materials like gold from some Hindu or Muslim sources. During the time of Abul Fazl impure gold was being assayed by the *salonī* process which involved firing with saltpetre and brickdust.

Chapter 2 in this monograph contains a detailed description about gem-minerals in pre-modern India. India was famous for diamond mining during the medieval period. There are occasional reports regarding new spurts of diamond mining activity in modern India. For example, the dried up beds of the Kṛṣṇā and Pennār river valleys, the kimberlite rocks in the Vijayawāḍā region, the village of Liṅgalapuṭṭa in the Viśākhāpaṭnam district — all in Andhra Pradesh — have shown fresh reserves of diamond deposits. The Government is contemplating to lease out the entire Cuḍḍapah basin to multinational companies for diamond mining. Similarly, the banks of the Mahānadī in the southern Orissa has been recently found to be 'virtual treasure trove of precious and semi-precious stones like aquamarine'.

Non-gem minerals in pre-modern India have been reviewed in chapter 3. It is worth emphasising that a few rock mineral names originated in the context of pre-modern India. Buchanan-Hamilton coined the term *laterite* for the brick-like ('lateritis' means brick) clayey iron-bearing red rock of Kerala. The famous *sehtā* of Rajasthan was found to contain the mineral *seypoorite* or *jaipurite* which has turned out to be $Co_{0.8}Fe_{0.2}AsS$ in chemical composition. *Mysorin* originally supposed to be a new mineral was eventually proved to be impure malachite. 'Dog metal' kyanite was discovered first in Orissa, India.

During the medieval period, several bio-geological deposits were processed in India and these have been discussed in chapter 3. Saltpetre contained potassium nitrate (used in explosives) and sodium nitrate (used for cooling water). Bacterial attack on sulphide mine deposits resulted in sulphates of iron, aluminium and copper, all subsequently leached, crystallised and used as mordants. In medieval China, iron powder was used to treat blue mine water containing copper sulphate to produce cementation copper. *Dhātukriyā* or *Dhātumañjarī* (71) described the chemical reaction of copper and mineral acids producing copper salts.

In his survey of the Indian iatrochemical period (AD 1300-1550), Ray has mentioned[3] specific names of minerals, inorganic compounds prepared from minerals, and of specific alloys. *Śārangadhara-saṁgraha* dated AD 1363 mentioned brass and bell-metal as specific *upadhātu* or alloys. *Madanapāla Nirghaṇṭu* (AD 1374) referred to *yaśadam vaṅgasadṛśam* or 'zinc, which is tin-like'. *Rasa-Ratna-Samuccaya* described the details of zinc metallurgy (chapter 5).

The final phase of the Hindu Chemistry developed during the end of the sixteenth century AD i.e., Akbar's reign. The Portuguese had established themselves fairly well at Goa and brought the dreadful venereal disease to India. The Tāntric antidote developed included *chobchini* or China root and *rasakarpūra* or calomel. The various iatrochemical texts of this period such as *Rasapradīpa, Rudrayāmala Tantra, Bhāvaprakāśa*, etc., mentioned the use of mercury salts and *śaṁkhadrāvaka* or the mineral acids obtained by distillation. These mineral acids, probably HCl, HNO_3 and H_2SO_3, H_2SO_4 solutions, were used as solvent for metals. The use was probably unknown in India before this time.

Rudrayāmala Tantra mentioned Ruma, the Arabic name for Constantinople and *dāhajala* (burning water) denoting sulphuric acid. *Arkaprakāśa* mentioned tinned copper vessel. The very title of the book was evidently Sanskritised from the Persian *arrak* (essence). Thus, it could be said that the Hindu alchemical tradition *started* interacting with the Muslim world rather late during the mature Moghul period.

Abul Fazl wrote about 'four elements' and seven categories of metals. Quicksilver was named as 'the mother of the bodies', and sulphur 'the father of the bodies'.[2] The mercury- sulphur theory probably represented the meeting point of the Chinese, the Muslim and the Hindu alchemical thoughts. The Hindu *Rasaśāstra* literature had earlier coined the term *sattva* for the *metallic essence* contained in the minerals. The Persian words were *ruh* and *arrak.*

Rudimentary concepts of metallic elements were implicit in the *Rasaśāstra* texts. The twelfth-century text *Rasārṇava* (VII.12-13, 20-1 and 41-4) described extraction of the *same* metal from the diverse sources of copper pyrites and copper sulphate: the same red copper resembling coccinella insect—*indragopasaṁkāśam sattvam*. The ninth-century text *Dhātuvāda*, later translated into Tibetan, described (*DV*, 2-3) precipitation of copper from copper salt solution using iron powder; the fifteenth-sixteenth-century AD text *Dhātukriyā* or *Dhātumañjarī* (*DKM*) described the reverse process: *tāmra dāhajalaiḥ yoge jāyate tutthakaṁ śubham* — copper when reacted with sulphuric acid produces copper sulphate (*DKM* 71).

Abul Fazl described the compositions of different kinds of brass and other alloys and narrated techniques for assaying silver-gold systems. The medieval India witnessed emergence of several unit operations and processess in the area of minerals and metals: leaching or lixiviation, as in the case of various sulphates, their evaporation and crystallisation, fractional or selective crystallisation of potassium nitrate from the saltpetre liquor, crushing, grinding and gravity separation of mineral systems, refinements in the crucible steel preparation, diverse alloy preparation in the copper-zinc-tin-lead phase systems, use of mercury in gold-amalgamation/purification systems and so on. These were often extensions of the earlier *Rasaśāstra* concepts. The categorisation of the iron-carbon alloys, the details of zinc metallurgy, etc., were unknown to Muslim scholars but vividly described in the thirteenth-century AD text *Rasa-Ratna-Samuccaya*. Eventually, there was the unique technology transfer — related to zinc distillation — from Rajasthan, India to England during the early eighteenth-century (chapter 5). On the other hand, certain foreign influences had flowed into the Indian iron and steel industries during the medieval period: this aspect previously neglected by the researchers has been emphasised in our chapter 4.

Iron and steel making in pre-modern India has been critically reviewed in chapter 4. The most salient feature in our critical and up-to-date review is the diversity of ore bodies treated, mineral beneficiation techniques and smelting patterns in the southern, central, eastern and western India. The recent replication experiments and theoretical studies have thrown considerable light onto these pre-modern smelting and crucible steel-making techniques.

We have reviewed different types of enterprises, the charcoal crisis and the various factors which led to stagnation and atrophy in the iron industries of pre-modern India. There were many noticeable foreign influences, but even these were not enough in quantity or quality to prevent the atrophy which could not be overcome without an adequate input of modern scientific knowledge.

In chapter 6, related to metallic art traditions in pre-modern India developed 'without steam or machinery', we have highlighted the then existing arts of enamelling, encrustation, jewel-settling, brass and Bidri wares, tin-coating of metallic utensils, gold and silver wares, etc. Many of these traditions have survived the ravages of time. As a matter of fact, some of the art traditions such as diamond and gold-based jewelleries are being rejuvenated.

We note from a recent report[4] that India consumed in 1993 some 424 tonnes of gold which was 17.4 per cent of the entire world production, and the highest consumption by any country in the world. India is currently exporting jewellery worth US $ 350 million, and yet 'the figure could increase by ten times if India's potential is properly tapped'. To help meeting this objective, the Geneva-based World Gold Council has decided to help in the modernisation and globalisation of the Indian jewellery designs. The Council will also focus on two areas — casting technology as well as electroforming, 'a process that will dominate jewellery manufacturing for years to come'.[4]

Why did the indigenous industries collapse?

This question has been handled in different chapters of our monograph, and it has been explained that there were many complex factors.

Firstly, there were several technical factors such as the charcoal crisis which had earlier bedevilled the Western world also. There was a certain degree of global inevitability in the charcoal-to-coal transition in ferrous industries and the accompanying problems of transition; India could not escape from this global trend. Similarly, the problems related to water-logged mines, their drainage, lack of mechanisation, etc., which crippled pre-modern mining and metallurgical industries in India, had earlier been faced and solved by Westerners.

We have mentioned following details in our previous chapters. In 1840, Ouseley discussed how the backdated Indian processes for gold and diamond recovery from riverbed sands could be mechanised and modernised (chapter 2, 32-4). Franklin had recommended in 1827 certain improvements in the mining of diamonds at Majgoha (chapter 2, 36-7). The various causes underlying the decline of the Indian diamond trade have been evaluated by Ball and Dutt (chapter 2, 38).

The editor of the *Journal of the Asiatic Society of Bengal* commented in 1838 that the mercury amalgamation-salt roasting method for purification of gold indigenously practised in India could be supplemented by galvanic magnetic method of beneficiation (chapter 3, 83). Dixon observed in 1831 the pre-modern tortuous lead mines of Ajmer and recommended modernisation through construction of shafts, galleries and adits (chapter 3, 79). Only a few machineries had come to India during eighteenth century: rolling and pilon mills, iron cylinders for grinding saltpetre, copper and iron boilers, and those too were brought by the Western merchants after the 1740s. The Indian rulers took little notice of these 'gadgets'.

Tapan Raychaudhuri presented a brief but critical and incisive analysis regarding the decadence of Indian industries during the pre-modern period.[5]

> Both early modern Europe and medieval China were far ahead of mid-eighteenth century India in such crucial fields of technology as the use of wind and water power, metallurgy, printing, nautical instruments, and basic tools and precision instruments.... The pattern was not one of total stagnation, but rather of a general indifference to labour saving devices.[5]

It is not that medieval India was unaware of the scientific and technological developments taking place in the West and in China. The Indian rulers were aware of clocks, print press, nautical instruments, telescope, various mechanical gadgets, wheel barrows, simple pumps and so on. Yet their response in favour of indigenous adaptation was strangely apathetic.[6-9] Why the apathy is a deeper question meriting deeper analysis.

Common artisans improved their techniques of iron-making, ship-building, etc. whenever and wherever they came in contact with the foreign experts such as in Sūrat, the Coromandel Coast, etc., but the foreigners, particularly the Dutch, were usually secretive, and the rulers of the country never appreciated the forces of the emerging modern science and technology.

The intellectual isolation was manifold. Debiprasad Chattopadhyaya mentioned evils of casteism which had ruined the glorious Hindu science in ancient India and continued to haunt medieval India. It haunts us even to-day.[10] The major impact of casteism has resulted in over-secretiveness in hereditary professions related to philosophical and practical knowledge. This kind of internal intellectual isolationism haunted the Muslim world in India as well. Ever since the days of Mahmūd Ghaznī, the Muslim invaders were hopelessly isolated from their own scientific and intellectual thought-world which had flourished in the countries of the Islamic tradition during the eighth-twelfth centuries AD. The Moghuls did not cultivate the earlier maritime adventurous spirit of Arabs and Persians. The eminent historian Irfan Habib has not adequately focused on this theme while interpreting medieval Indian history.[11]

Discussing the intrusion of Islam in India, Irfan Habib has categorised his predecessor historians as either 'communal' or 'secular' and belittled the works of H.M. Elliot and R.C. Majumdar.[12] Habib could not however question the veracity of the facts presented by Elliot and Majumdar or the original accounts of Bernier.

> The country (Mughal period) is ruined by the necessity of defraying the enormous charges required to maintain the splendour of a numerous court, and to pay a large army maintained for the purpose of keeping the people in subjection (majority Hindus). . . . A profound and universal ignorance is the natural consequence of such a state of society as I have endeavoured to describe. Is it possible to establish in Hindustan academies and colleges properly endowed? — Bernier's *Travels*

Irfan Habib could not contest the fact that the Hindu subjects remained bitter against the Muslim rulers and the intellectual isolation between the two communities seriously affected the flow of scientific knowledge within India. This could have been one of the reasons why scientific revolution did not take place in India.[9] Ray Chaudhuri has also accepted this point.[13]

Nobody holds any brief in favour of the despotic British rulers, who were no better than the earlier Muslim rulers of the country. Occasionally we got one Ripon like the exceptional, Akbar. The major difference between the two eras — the Musilim and the British — has been that the latter tyrants brought into India (may be inadvertently) modern science. This fact was sportingly appreciated by Rammohun Roy, Tagore and Vivekānanda but not by Mahatma Gandhi, Jawaharlal Nehru[14] and Dharampal.[15] Recently Sen[16] and Blanpied[17] have restored the balance of logical analysis supporting Rammohun Roy. As a matter of fact, the British introduced modern science into India too slowly and in a niggardly fashion. Thus, Dharampal's view that Western science was thrust upon India and too quickly cannot be accepted. During the eighteenth/nineteenth centuries, the Japanese had sportingly and wisely abandoned their Chinese moorings and learnt modern science from the Western colonial powers.

What we accept from Dharampal's thesis is the fact of brutal economic and political exploitation of the Indian people by the British starting from the eighteenth century. But then a similar, if not worse, period of brutality had started six centuries earlier; some historians would like to hide this unpalatable fact under the carpet of 'secularism'![18]

Where we are in perfect union of thoughts with Dharampal is the legitimate pride in the indigenous industries and traditions in pre-modern India. Here the bulk of the credit goes to the indomitable artisans representing so many castes, religions, languages and sub-cultures. They exhibited gallant and persistent craftmanship. Their vitality was sustained through centuries and could not be extinguished either by the Hindu obscurantism, Muslim fanaticism or the British colonialism. Svāmī Vivekānanda saluted the working classes in his beautiful language:

> Those uncared for lower classes of India — the peasants and weavers and the rest, who have been conquered by foreigners and are looked down upon by their own people — it is they who from time immemorial have been working silently, without even getting the remuneration of their labours!
>
> Who cares to praise those whose heart's blood has contributed to all the progress that has been made in the world? . . . Our proletariat are doing their duty in their homes day and night, without the slightest murmur — well, is there no heroism in this? . . . Ye ever-trampled labouring class of India ! I bow to you.[19]

We conclude this chapter as well as this monograph by stressing the utility of the present studies. Some of the pre-modern era indigenous technologies need to be studied carefully for possible revival. Low height direct reduction furnaces had no doubt low productivity and low thermal efficiency. But their impurity pick-up was less and the metal produced was more malleable. The pre-modern technique of direct smelting of iron ore may be revived with certain modifications. Rice-husk reinforced refractory crucibles can still be tried for high temperature work. The ancient art of making wootz steel and Damascus blade can be simulated to make ultra-high carbon steel having strength as well as ductility. The glorious tradition of producing art objects even 'without steam or machinery' can be pursued and, as World Gold Council and the diamond-based industries have shown, improved upon through the use of machineries and new adaptations.

History of science studies have their own utilities like all historical studies. The present investigations of minerals, gems and metals in ancient and pre-modern India have served two purposes. Firstly, the glories of the past material traditions of our nation have been documented and the sources of internal vitality pin-pointed. Secondly, attempts have been made to identify reasons for stagnation and eventual decay of the indigenous industries based on minerals, gems and metals. The obvious lesson is that our national deficiencies should be rectified so that the mistakes committed in the past are not repeated. Critical and utilitarian studies such as ours should be pursued.

References

1. A.K. Biswas, *Minerals and Metals in Ancient India*, Two Volumes, Report on the Sponsored Work (1987-1991) submitted to the Indian National Science Academy, New Delhi, Vol. I, Travellers'Accounts and Archaeological Evidences, 21 chapters, 59 tables and 24 figures. Vol. II, Literary Evidences. 10 chapters, 3 tables and 5 appendices. D.K. Printworld Pvt. Ltd., New Delhi has published the entire material in 2 volumes.
2. Abul Fazl, *The Āin-ī-Akbarī*, Vol. I. tr. by H. Blochmann, New Delhi, 3rd edition, 1977, pp. 40-5.
3. P. Ray (ed.), *History of Chemistry in Ancient and Medieval India*, incorporating the History of Hindu Chemistry, Indian Chemical Society, Calcutta, 1956, pp. 158-64.
4. Ranvir Nayar, A Golden Opportunity, *SUNDAY Magazine*, 19-25 June 1994, pp. 62-3.
5. Tapan Raychaudhuri, the Mid-Eighteenth Century Background, in *The Cambridge Economic History of India*, Vol. II ed. by Dharma Kumar, Orient Longman and Cambridge University Press, 1982, pp. 18-21 and 31-5.
6. A.J. Qaisar, *The Indian Response to European Technology and Culture* (AD 1498-1707), Oxford University Press, 1982.
7. Satpal Sangwan, Indian Response to European Science and Technology (1757-1857), *British Journal of History of Science, 21*, 1988, pp. 211-32.
8. Satpal Sangwan, *Science Technology and Colonisation (1757-1857)*, Anamika Prakashan, New Delhi, 1991.
9. Satpal Sangwan, Why did the Scientific Revolution not take place in India? in *Science and Empire — Essays in Indian Context*, ed. by Deepak Kumar, Anamika Prakashan, New Delhi, 1991, pp. 31-40.

10. Debiprasad Chattopadhyaya, *History of Science and Technology in Ancient India*, Vol. I. The Beginnings, Firma KLM Pvt. Ltd., Calcutta, 1986; Vol. II, Formation of the Theoretical Fundamentals of Natural Science, 1991.

11. Irfan Habib, *Interpreting Indian History*, North-Eastern Hill University Publications, Shillong, 1985.

12. *Ibid*., pp. 29-33.

13. Ref. No. 5, p. 31.

14. Jawaharlal Nehru, *The Discovery of India*, Anchor Books, Doubleday and Company, Inc., p. 197.

15. Dharampal, *Indian Science and Technology in the Eighteenth Century*, Academy of Gandhian Studies, Hyderabad, 1971, Reprint 1983, pp. 17-58.

16. S.N. Sen, The Character of the Introduction of Western Science in India During the Eighteenth and Nineteenth Centuries, *Indian Journal of History of Science 1*, No. 2, 1966, pp. 112-22.

17. W.A. Blanpied, Notes for a Study on the Early Scientific Work of the Asiatic Society of Bengal, *Japanese Studies in the History of Science*, No. 12, 1973, pp. 121-44.

18. Arun Kumar Biswas, Minerals and Metals in Medieval India, chapter 11, pp. 275-313 in *History of Indian Science, Technology and Culture* (AD 1000-1800) edited by A. Rahman (Vol. III, Part I of a PHISPC Publication series), Oxford University Press, New Delhi, 2000.

    This paper contains our detailed analysis on the nature of medieval decadence in Indian science and technology. Specially note pp. 275-7 and 307-10.

19. Swami Vivekananda, Memoirs of European Travel, *The Complete Works*, Mayavati Memorial Edition, Vol. VII, Advaita Ashrama, Calcutta, 1979, pp. 358-9.

A.1. Great Mogul 280 carats

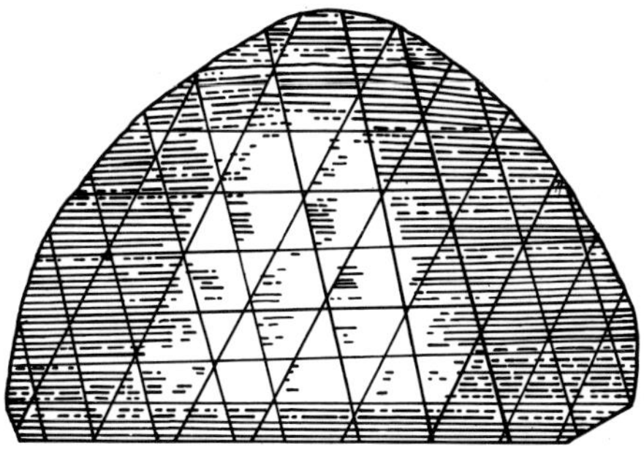

A.2. Citron color 140 carats

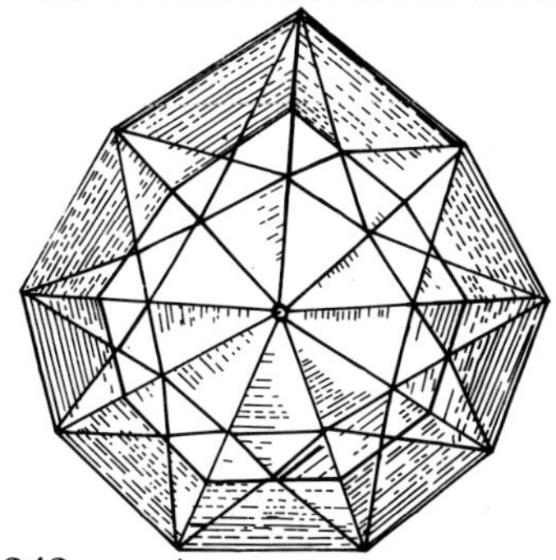

A.3. Golconda 1642 A.D. 242 carats

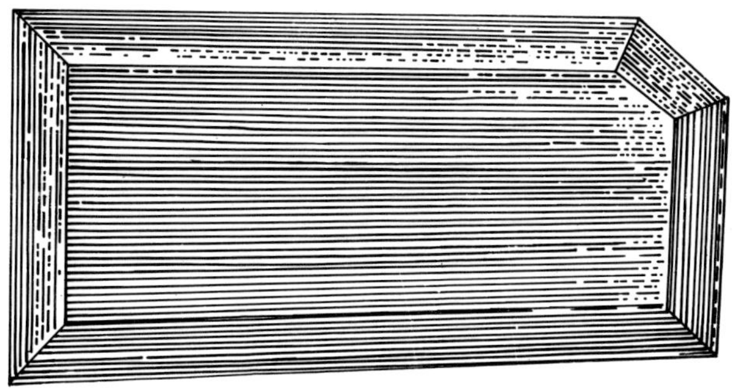

A.4. Ahmedabad 157 carats

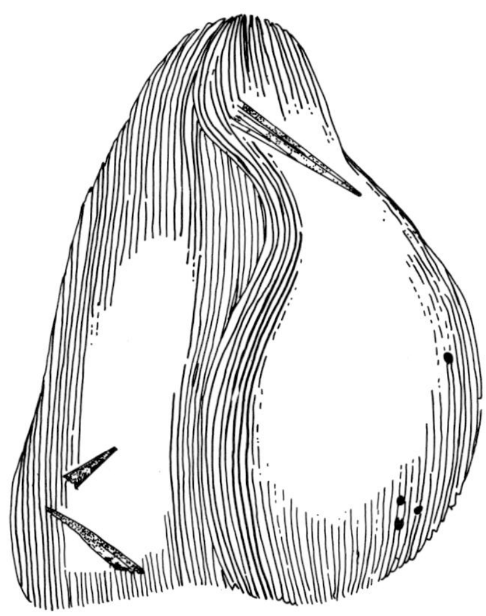

A.5. 94 carats

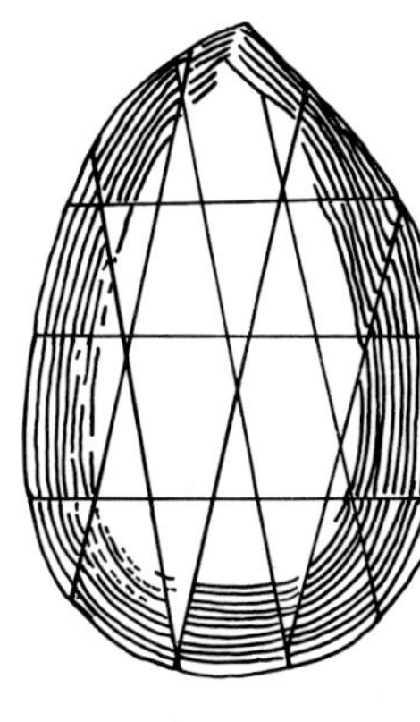

A.6. Kollur Mine 64 carats

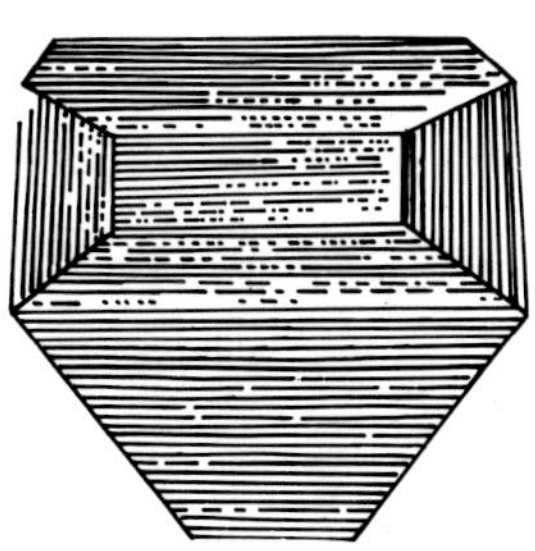

A.7

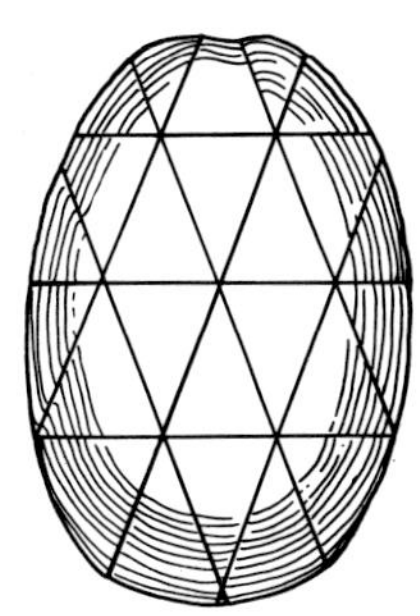

A.8

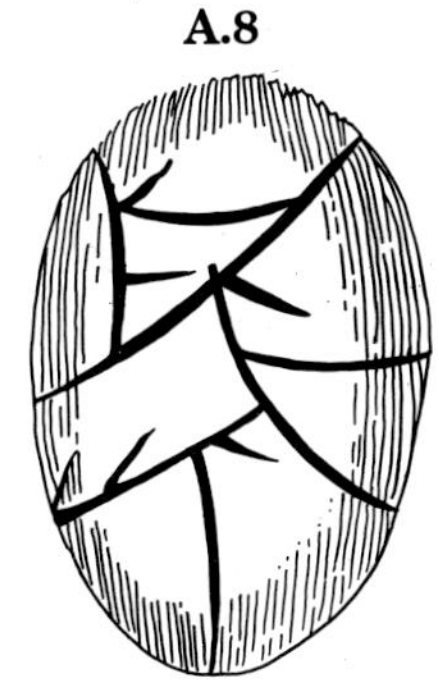

Pl. 1: Figures of large diamond — Tavernier

Pl. 2: Diamond mining near Pannā in the year 1867

Pl. 3 : Gold Washing

Pl. 4 : Dish and scrapes for gold washing

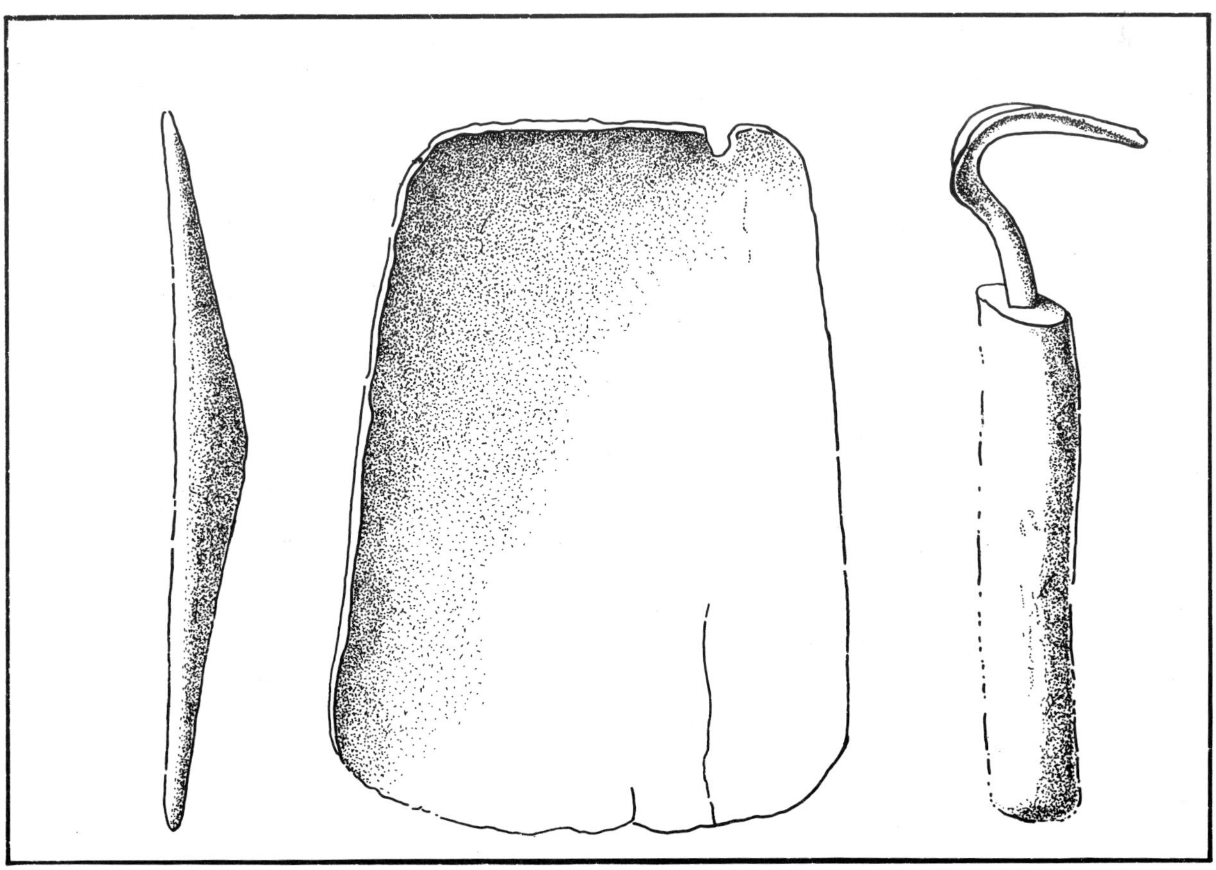

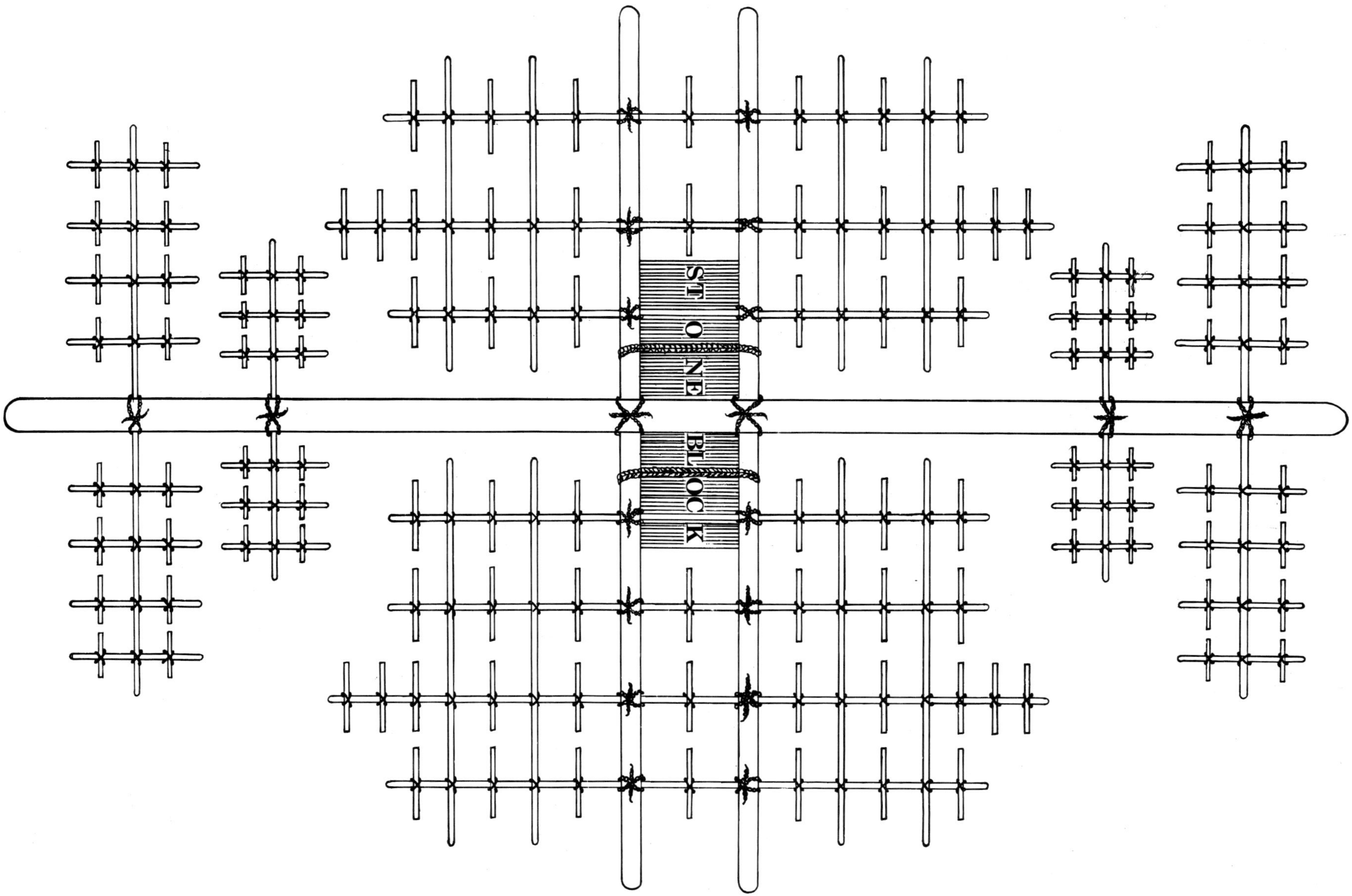

Pl. 5: Frames used for moving large stones

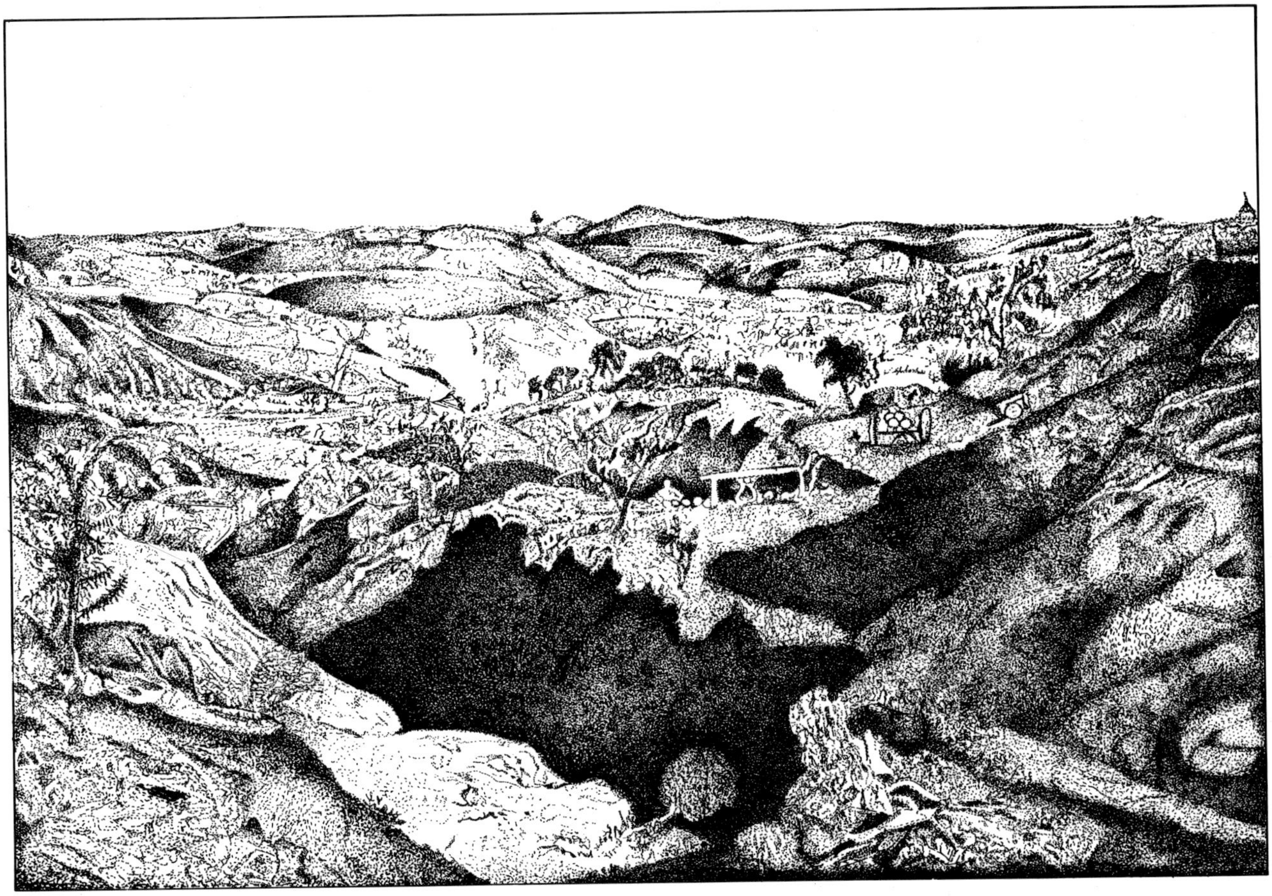

Pl. 6 : Petroleum wells

Pl. 7: Gin and coal-pit head

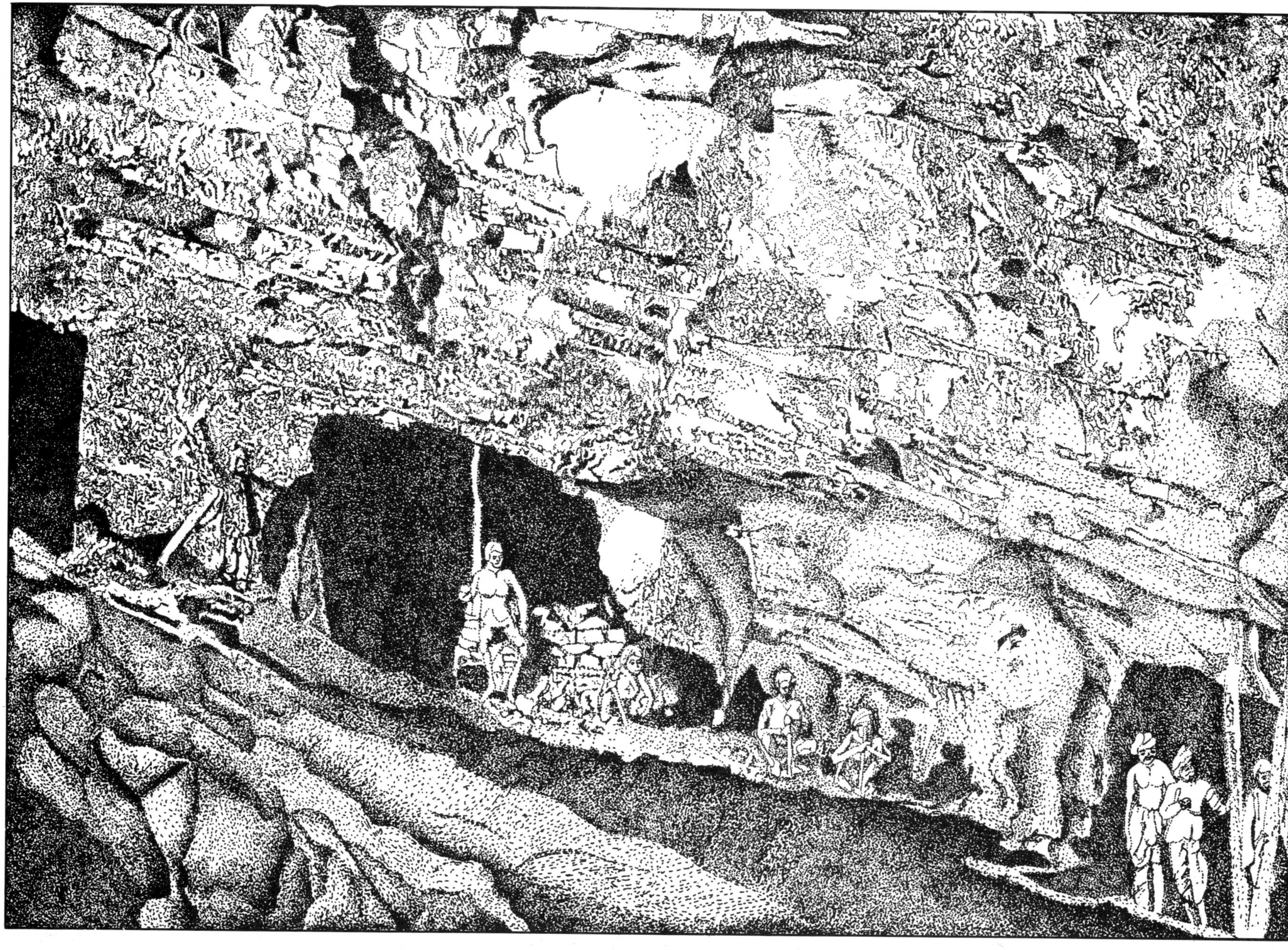

Pl. 8: Entrance to Mopani Coal Mine

Pl. 9: Hydraulic mining, Khasi Hills

Pl. 10: Agaria iron smelting

Pl. 11: Eighteenth-nineteenth century 94.5 cm. long sword of the Maharaja of Holkar set with rubies, emeralds, diamond and onyx, Victoria & Albert Museum, London.

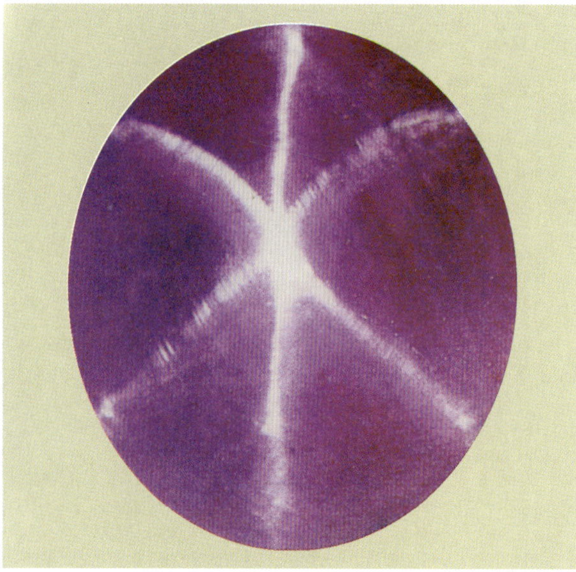

Pl. 12: Rosser Reeves Ruby (138.7 carats) from Sri Lanka — the largest and finest star ruby in the world.

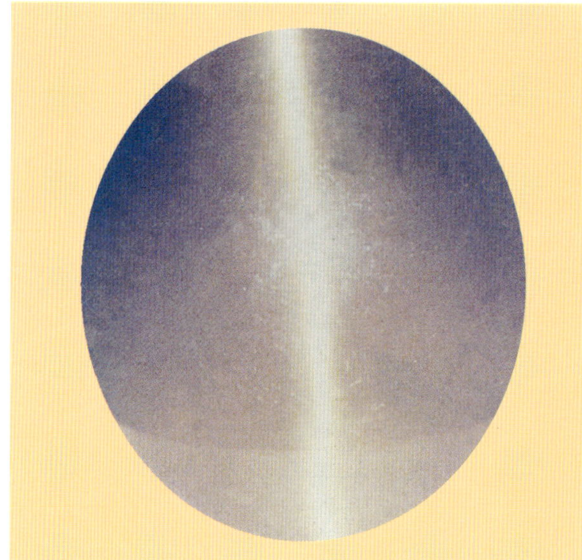

Pl. 13: The Maharani — 58.2 carat chrysoberyl cat's eye.

Pl. 14: The Taj Mahal Emerald, Mughal, *c.* 1630-50, Colombian emerald, 4 x 5 cm. weight: 141.13 carats.

Pl. 15: Panning for gems in the Mahānadī.
Inset an aquamarine or probably green beryl.

Pl. 16: 911 carat aquamarine from Brazil.

Pl. 17: Weight for Floor Covering, Mughal, *c.* 1700, alabaster with gemstone inlay, 9.5 x 8.5 cm.

Pl. 18: Emerald. The name was derived from the Sanskrit word *Marakata* meaning the (Egyptian) gem found near desert (*maru*) and sea-coast (*kata*), Masāraka (Egyptian gem) and Smaragdos.

Pl. 19: Istanbul Dagger set with three emeralds, Istanbul Museum.

Pl. 20: General view of Zawar

Pl. 21: Zawar Temple

Pl. 22: Retort walls, Zawar

Pl. 23: Rāṇā Pratāp Khan-mine, Zawar

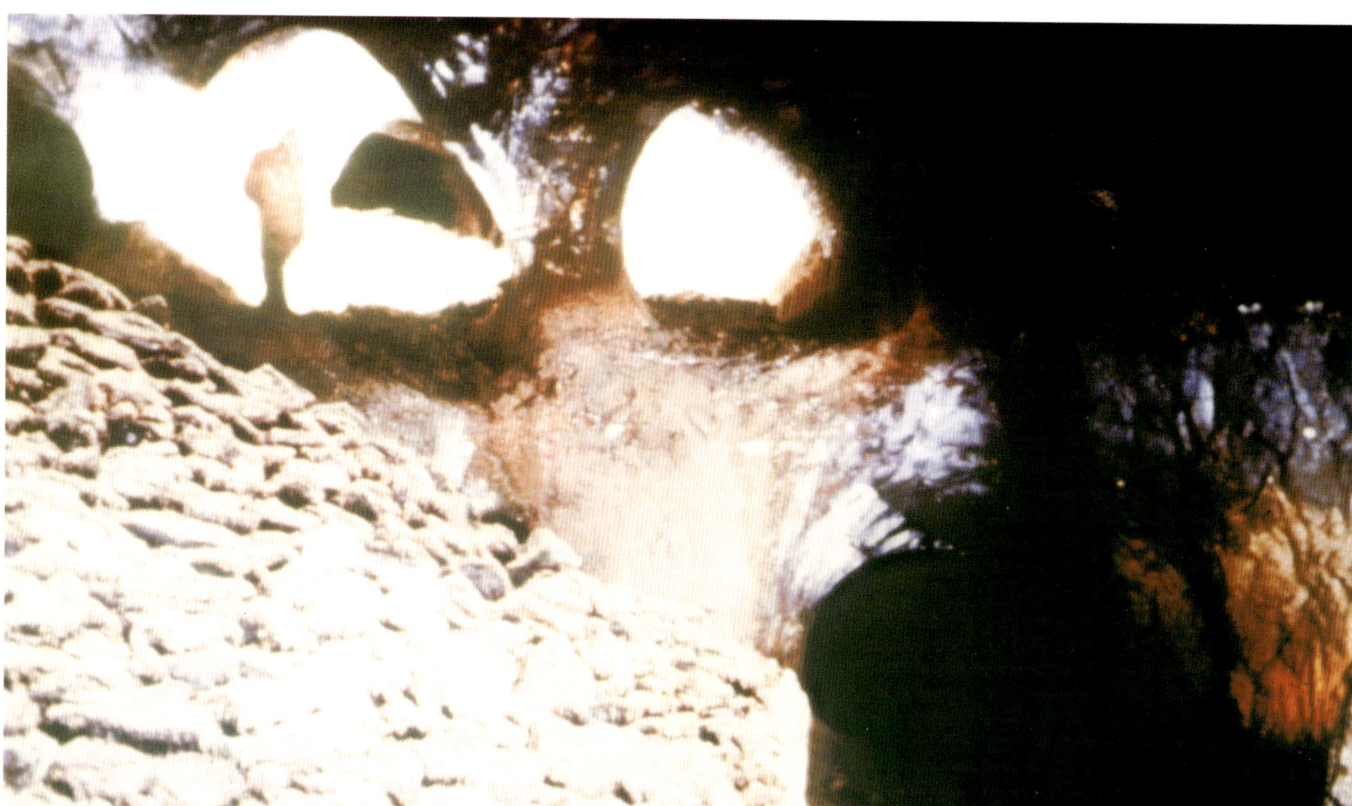

Pl. 24: Slag heaps, Zawar

Pl. 25: Retort plate, Zawar

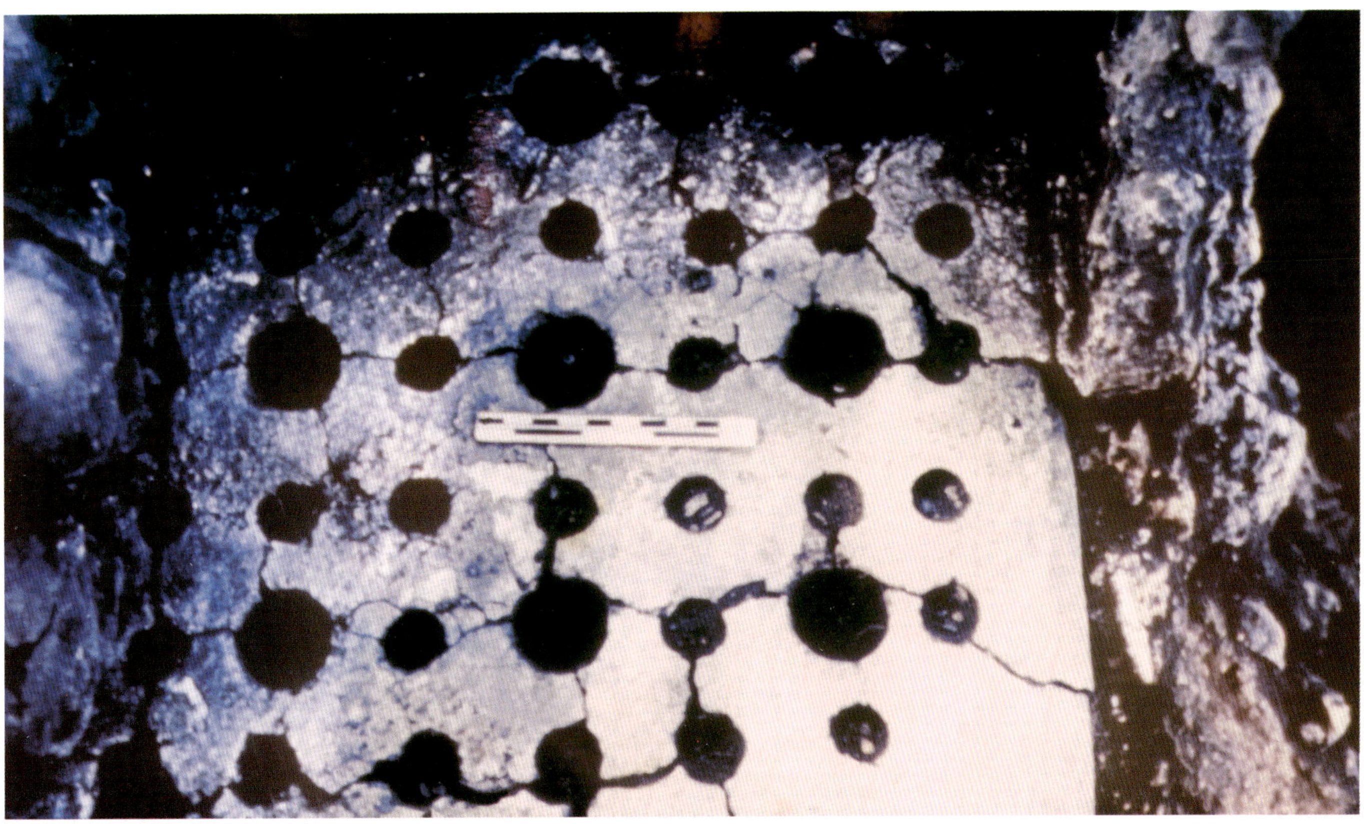

Pl. 26: Retort bank I, Zawar

Pl. 27: Retort bank II, Zawar

Pl. 28: Underground mine, Haulage, Zawar

Pl. 29: Model of furnace — artist's view, Zawar

Pl. 30: Alum and sulphate of copper works at Khetri

Pl. 31: Malik-i-Maidan bronze gun at Bijapur

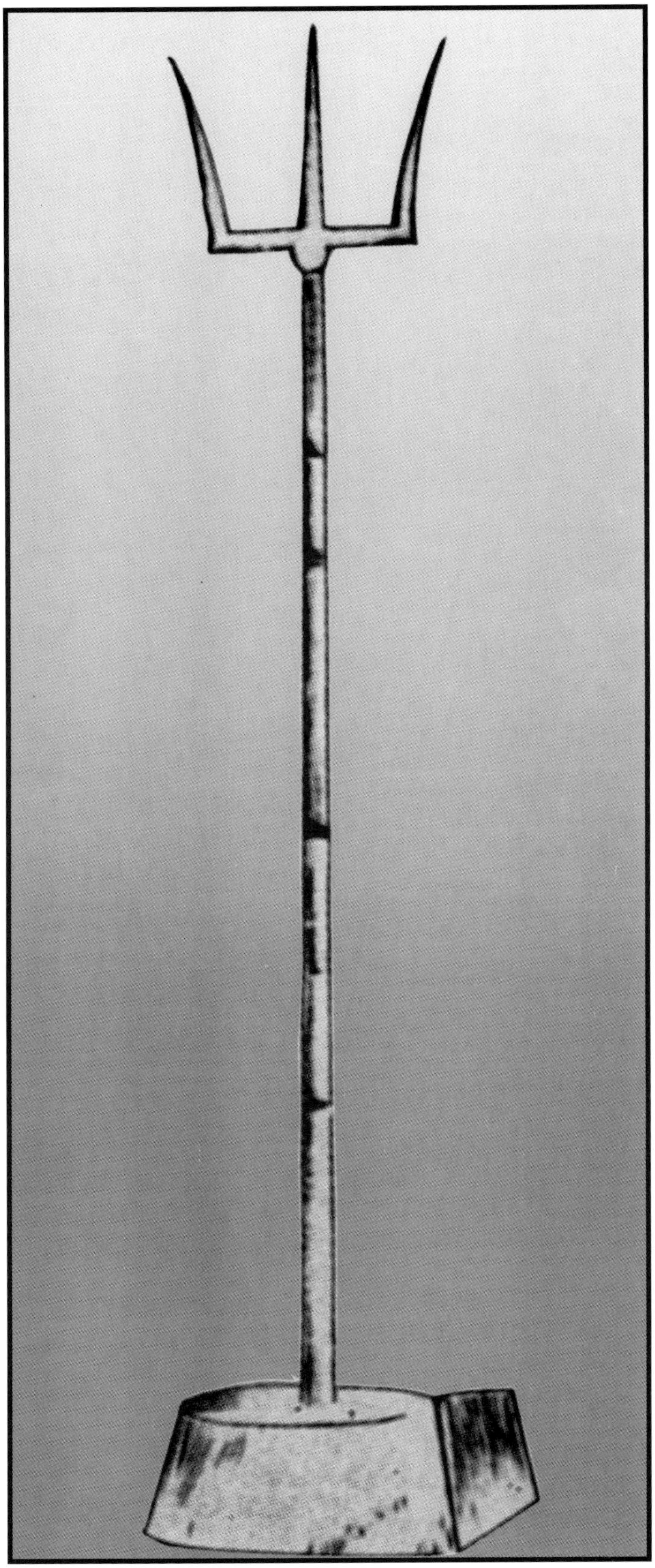

Pl. 32: Iron pillar, Mount Abu

Pl. 33: Iron pillar, Dhar

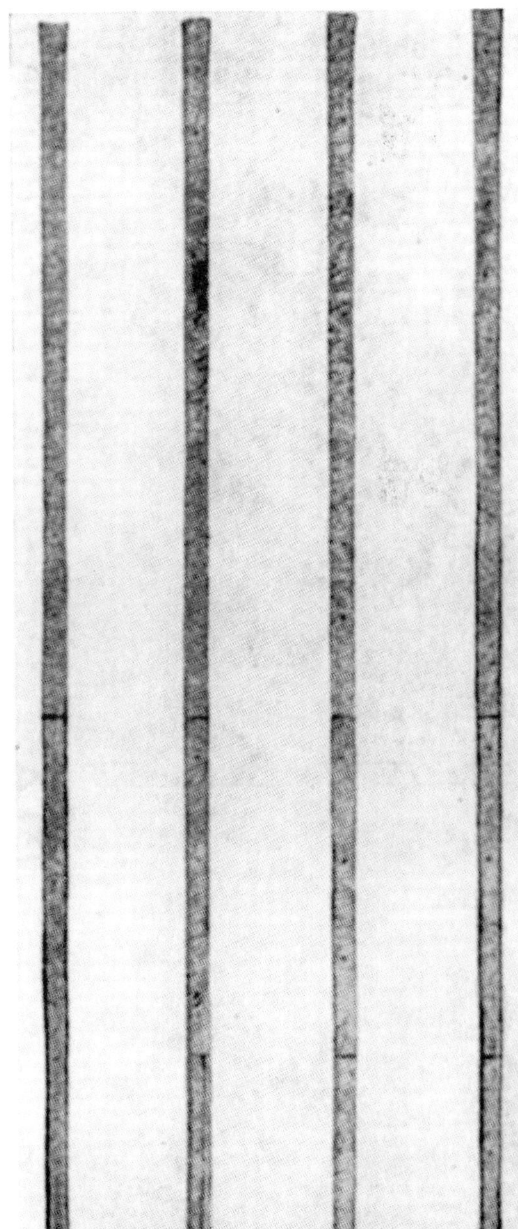

Pl. 34: Huqqa base, a Bidriware inlaid with silver and brass. Second half of seventeenth century.
Height 18.6 cm, Diam. 16.8 cm. Silver has been used for the flowers and brass for the leaves and stems.

Pl. 35: Ewer, A Bidriware, inlaid with silver and brass, inscribed on the rim.
The base metal is high-zinc-low-copper alloy. The black patina is due to KNO_3
— oxidation of copper-rich phase. The design is cut out of sheet silver with brass wire being used to indicate details.
The inscription suggests that the ewer was made for Timur in 1406 or 1407. Height 28.5 cm, width 18.5 cm.

Pl. 36: Iron pillar, Konark

Pl. 37: Iron gun, Gulbarga

Pl. 38: Copper reduction at Khetri

Pl. 39: Vr̥ntāka-Mūṣā, brinjal-like Zinc (ore) reduction-distillation crucible from Zawar.
Slag in broken Zinc distillation retorts.

Bibliography

Agarwal, D.P., 1984, *The Archaeology of India*, Select Book Service Syndicate, New Delhi.

Bag, A.K. (ed.), 1997, *History of Technology in India*, vol. I, Indian National Ṡcience Academy, New Delhi. This volume contains two papers by A.K. Biswas on gems and minerals and social factors.

Ball, V., 1881, *A Manual of the Geology of India*, Part III, Economic Geology, GSI, Calcutta.

Bauer, Max, 1968, *Precious Stones*, 2 vols., Dover Publications, ed. and tr. of *Edelstein Kunde*, 1896.

Biswas, Arun Kumar, 1969, *Science of India*, Firma KLM Private Ltd., Calcutta.

———, 1996, *Minerals and Metals in Ancient India*, vol.I: Archaeological Evidence; vol. II: Literary Evidence (co-author: Sulekha Biswas).

——— (ed.), 2001, *History, Science and Society in the Indian Context*, The Asiatic Society, Calcutta.

Blochmann, H., 1977, *The Āin-i-Akbarī of Abul Fazl* (ed., tr. and rev. by D.C. Phillott), Oriental Books Reprint Corporation, New Delhi.

Bose, D.M., Sen,S.M. and Subbarayappa, B.V., 1971, *A Concise History of Science in India*, Indian National Science Academy, New Delhi.

Bose, Pramatha Nath, 1975, *A History of Hindu Civilisation during British Rule*, 1894, 3 vols., rpt., Asian Publication Services, New Delhi.

Brown, C.J. and Dey, A.K., 1955, *India's Mineral Wealth*, Oxford University Press, Bombay.

Chattopadhyaya, Debiprasad, 1986 and 1991, *History of Science and Technology in Ancient India*, vol. I: The Beginning and vol. II: Formation of the Theoretical Foundations of National Science, Firma KLM Private Ltd., Calcutta.

Coomaraswamy, A.K., 1956, *Medieval Sinhalese Art*, Pantheon, New York.

Craddock, P.T. and Hughes, M.J. (eds.), 1985, *Furnaces and Smelting Technology in Antiquity*, Occassional Paper No. 48, British Museum, London.

Craddock, P.T. (ed.), 1990, *2000 Years of Zinc and Brass*, Occassional Paper No. 50, British Museum, London.

Dahiya, Neelima, 1986, *Arts and Crafts in Northern India*, B.R. Publishing Corporation, Delhi.

Dana, E.S., 1961, *A Textbook of Mineralogy*, John Wiley and Sons, New York.

Dharampal, 1983, *Indian Science and Technology in the Eighteenth Century*, Academy of Gandhian Studies, Hyderabad.

Dharma Kumar, 1982, *The Cambridge Economic History of India*, vol. II (AD 1757-1970), Cambridge University Press.

Elwin, Verier, 1942, *The Agaria*, Oxford University Press, Calcutta.

Gode, P.K., 1961, *Studies in Indian Cultural History*, Visveshvaranand Vedic Research Institute, Hoshiarpur.

Habib, Irfan, 1985, *Medieval Technology Exchanges between India and the Islamic World*, Aligarh Oriented Series No. 6, Viveka Publications, Aligarh.

———, 1985, *Interpreting Indian History*, North-Eastern Hill University Publications, Shillong.

Habib, Mohammad, 1937, *The Campaign of Alauddin Khilji*, by Amir Khusrau tr., Taraporevala and Sons, Bombay.

Hegde K.T.M., 1991, *An Introduction to Ancient Indian Metallurgy*, Geological Society of India, Bangalore.

Joshi, D. (ed. and tr.), 1987, *Rasaratnasamuccaya of Vāgabhaṭācārya*, Indian National Science Academy, New Delhi.

Joshi, S.D., 1970, *History of Metal Founding on the Indian Sub-Continent Since Ancient Times*, Ranchi.

Krishnan, M.S., 1955, *Iron Ore of India*, Calcutta.

Kuppuram, G. and Kurnudamani, K., 1990, *History of Science and Technology of India*, vol. 8, Sandeep Prakashan, New Delhi.

Mahmud, Sayed Jafar, 1988, *Metal Technology in Medieval India*, Daya Publishing House, New Delhi.

Mallet, F.R., 1887, *A Manual of the Geology of India*, Part IV, *Mineralogy*, Geological Survey of India, Calcutta.

Moreland, W.H., 1972, *From Akbar to Aurangazeb*, Oriental Books Reprint Corporation, New Delhi.

Nahta, Bhamarlal (ed.), 19976, *Dravyaparikkha aur Dhatu utpatti* by Ṭhakkura Pherū, Vaishali.

Neogi, Panchanan, 1918, *Copper in Ancient India*, Bulletin 12, Indian Association for Cultivation of Science, Calcutta.

———, 1940, *Iron in Ancient India,* Bulletin 12, Indian Association for Cultivation of Science, Calcutta.

Pal, M.K., 1983, *Metal Casting in South India*, Seminar on Ancient Metal Industries of South India, Thanjavur. The Proceedings contain many other papers such as Gold in Ancient South India by G. Kuppuram, etc.

Pal, Pratapaditya, 1975, *Bronzes of Kashmir*, Graz.

Percy, John, 1864, *Metallurgy of Iron and Steel*, London, U.K.

Qaisar, A.J., 1982, *The Indian Response to European Technology and Culture 1498-1707*, Oxford University Press, New Delhi.

Rahman, A., 1982, *Science and Technology in Medieval India — A Bibliography of Source Materials in Sanskrit, Arabic and Persian*, Indian National Science Academy, New Delhi.

——— (ed.), 2000, *History of Indian Science, Technology and Culture* AD *1000-1800*, vol. III, Part 1 of PHISPC Series, Oxford University Press, New Delhi. This volume contains an article *Minerals and Metals in Medieval India* (pp. 275-313) by A.K. Biswas in which the present author has made a detailed analysis of the pre-modern decline of Indian S & T.

Ray, Amita and Mukherjee, S. (eds.), 1990, *Historical Archaeology of India*, Books and Books, Delhi.

Ray, P. (ed.), 1956, *History of Chemistry in Ancient and Medieval India*, Indian Chemical Society, Calcutta.

Ray Chaudhuri, Tapan and Habib, Irfan, 1982, *The Cambridge Economic History of India*, vol. I (1200-1750), Cambridge University Press, rpt. in 1993.

Roy, Aniruddha and Bagchi, S.K. (eds.) 1986, *Technology in Ancient and Medieval India*, Sundeep Prakashan, Delhi. This volume contains many useful articles.

Sangwan, Satpal, 1991, *Science, Technology and Colonisation (1757-1857)*, Anamika Prakashan, New Delhi. Vide also his article: Why did the Scientific revolution not take place in India? in *Science and Empire — Essays in Indian Context*, ed. by Deepak Kumar, Anamika Prakashan, New Delhi, 1991.

Sarma, S.R., 1984, Ṭhakkura Pherū's *Rayaṇaparikkhā*, Viveka Publication, Aligarh.

Sarton, George, 1931, *Introduction to the History of Science*, Carnegie Institution of Washington, rpt. in 1962.

Sen, S.N. (ed.), 1986, *The Cultural Heritage of India, Volume Six on Indian Science*, Ramakrishna Mission Institute of Culture, Calcutta.

Smith, C.S., 1960, *A History of Metallurgy*, University of Chicago Press, Chicago.

Stronge, S., 1985, *Bidri Ware, Inlaid Metal-work from India*, Victoria and Albert Museum, London, U.K.

Subbarayappa, B.V. (ed.), 1985, *Indo-Soviet Seminar on Scientific and Technological Exchanges between India and Soviet Central Asia in Medieval Period*, Indian

National Science Academy, New Delhi, This volume contains many useful articles.

Swarna Kamal, 1980, *Studies in Metallic Art and Technology of Gujarat*, Museum & Picture Gallery, Baroda.

Tagore, S.M., 1879, *Maṇimālā*, 2 vol., Calcutta.

Tavernier, Jean Baptiste, 1676, *Travels in India* (original in French) translated in 1889 by Valentine Ball.

Tod, J., 1829, *Annals and Antiquities of Rajasthan*, Smith and Elder, London, 1950 edition published by Routledge and Kegan Paul Limited.

Tylecote, R.F., 1962, *Metallurgy in Archaeology*, British Museum and Edward Arnold, London.

———, 1976, *A History of Mettallurgy*, The Metal Society, London, 2nd edn. 1979 by Institute of Materials.

———, 1984, *The Metal and Material Technology*.

Vandiver, P. (ed.), 1990, *Materials Issues in Art and Archaeology: Proceeedings of the Symposium*, Materials Research Society. This volume contains articles on wootz, crucible refractory, etc.

Wiedemann, Eilhard, 1970, *Aufsaetze Zur Arabischen Wissenchaftsgeschichte*, Hildeshein, New York. This book in a good treatise on mineralogy in Islam era science.

Index